The Matter of Facts

The Matter of Facts

Skepticism, Persuasion, and Evidence in Science

Gareth Leng and Rhodri Ivor Leng

The MIT Press
Cambridge, Massachusetts
London, England

This book was set in Stone Serif and Stone Sans by Jen Jackowitz. Printed and bound in the United States of America.

Library of Congress Cataloging-in-Publication Data

Names: Leng, G. (Gareth), author. | Leng, Rhodri, author.
Title: The matter of facts : skepticism, persuasion, and evidence in
    science / Gareth Leng and Rhodri Ivor Leng.
Description: Cambridge, Massachusetts : The MIT Press, [2020] |
    Includes bibliographical references and index.
Identifiers: LCCN 2019029887 | ISBN 9780262043885 (hardcover)
Subjects: LCSH: Science--Methodology. | Evidence. | Empiricism. | Facts
    (Philosophy) | Research--Evaluation. | Communication in science. |
    Science publishing--Moral and ethical aspects.
Classification: LCC Q175.32.T78 L46 2020 | DDC 500--dc23
LC record available at https://lccn.loc.gov/2019029887

10   9   8   7   6   5   4   3   2   1

Contents

Preface

Rhodri:

This book grew from conversations with my father about what Science involves, what makes it special, and what causes it problems.

In 2016, we found ourselves working at the same university for the first time in our careers. Whereas he was approaching retirement after a long and distinguished career in experimental physiology, I was just beginning mine as a social scientist interested in the way evidence spreads in the scientific community. Despite our different backgrounds, we shared an interest in how science develops, how evidence is produced and used, and how the evaluation of evidence is taught at university.

This book is the result. It is not a systematic introduction to the sociology, philosophy, or history of science but about particular issues that we feel need reflection. We have tried to capture our admiration for the scientific process as well as an understanding of its complexity and fragility in the face of the many problems that now confront it. We don't propose answers to the problems we survey, but we do offer a context in which these problems may be understood. This book isn't trying to simplify science or direct you to a particular way of seeing it. It is there to provoke you.

Our hope is that this book might help students to see the contestable nature of knowledge, to get them to engage seriously with methodology, with empirical findings, with logic, and with the important flaws in Science that they must be aware of. For those who, like me, are at the beginning of their academic careers, I hope that you find it as useful as I have in navigating the complexities and pressures of academic life. I've been troubled by the prevalence of bias and poor research design, inspired by scientists confronting the dogma of their fields and pursuing novel research, and reassured that Science, though flawed in important ways, remains worth pursuing. Science is a human endeavor, and it is beset by the flaws all humans have, but it is endowed with their virtues too.

I need to first thank my partner, Hayley Urquhart, whose unwavering support and patience was tested to the limit with my decision to simultaneously write a book alongside my PhD.

I would also like to thank all the students that I have taught over the years at Edinburgh. Seeing you develop into confident researchers at ease with questioning and scrutinizing data and theory has instilled a lasting faith in me of the virtues of critical discussion and scholarship. I also wish to thank Daniel Kenealy, Lindsay Paterson, Richard Brodie, and Miguel Garcia-Sancho for giving me the opportunity to teach and for their valuable guidance. I must also thank Gil Viry for introducing me to network analysis and teaching me its fundamentals. Finally, I would like to thank my supervisors, Steve Sturdy and Valeria Skafida, for their support and understanding over the years, and the Economic and Social Research Council for providing me with funds to pursue a PhD through their Advanced Quantitative Methods scholarship.

Gareth:

As Rhodri says, this book sprang from conversations between a father and son, between a sociologist and a scientist, between a teacher and a student. I was more the student, and he perhaps more the scientist.

I must thank George Fink for putting me right on Geoffrey Harris, and Tony Trewavas and Steve Hillier for sharing their anecdotes. We must also thank another of my sons, Trystan, for painstakingly reading through the draft—he is also a network scientist, as well as being a philosopher and mathematician. And we must also thank Nancy Sabatier, whose sharp eyes caught some embarrassing blunders, and whose constant support was priceless.

My father, Rhodri's grandfather was Ivor John Leng, a sociologist himself, a pioneer in quantitative methods, and author of *Children in the Library* (University of Wales Press, 1968). He taught me what it means to be a scholar. If this book is about anything, it is about that, and we dedicate this book to his memory.

Rhodri Leng and Gareth Leng

June 2019

Prelude: Sources

At the foot of the Gerbier de Jonc, a gorse-covered rock in the Ardèche, a spout of clear water emerges from a wall within an old stone barn. Outside, a sign declares *ici FERME de la LOIRE—SOURCE GEOGRAPHIC de la LOIRE*. Here, it seems, is the source of the Loire, the longest river in France, that glorious river celebrated in paintings by Monet and Sisley, a river once navigated by Vikings to sack the city of Tours, a river which, during the Hundred Years War, marked the border of English possessions in France.

In what sense is this trickle the source of the Loire? If we stem it, will children in Tours go thirsty? Across the road, another sign simply declares "La Loire"—this is *La Source Authentique*. From day to day, this spring may flow or not—it depends on the level of the water table. About a kilometer south is *La Source Véritable*—the "true source"—where a hand-written sign declares *ici commence ma longue course vers l'Océan*. This spring flows into a trout pond. Close by is a marsh, fed by a hidden spring whose waters tumble into a wooded ravine. Down the valley, the waters from these springs converge to become a stream that merges with other streams to become a rivulet fed by other tributaries to become the great river that winds through villages, towns, and vineyards. In the nineteenth century, before railways were built, this was a thoroughfare that each year carried tens of thousands of passengers in steamboats between Nantes and Orléans.

But when two waters merge, which is the river and which the tributary? This depends on which flow is the greater; this can be hard to measure, and each flow will vary with season and rainfall. The largest tributary of the Loire, the Allier, rises in the Massif Central and joins the Loire just south of Nevers where, at their confluence, the Allier is said by some to run *stronger* than the Loire, in which case perhaps we should be talking about the Châteaux and vineyards of the Allier Valley, rather than of the Loire.

The Loire runs north from the Ardèche before turning west to meet the Atlantic at Saint-Nazaire. But during the Pleistocene, the "paleo-Loire" continued to flow north and joined the Seine, while the lower Loire began near Orléans and flowed west along the present course of the Loire. At some time during the history of uplift in the Paris Basin, the lower Loire captured the palaeo-Loire producing the present river. The Loire today is not the Loire as it once was.

Every guidebook gives the length of the Loire as 1,012 kilometers. How was this measured? The Loire nowhere runs straight; does this length track every bend?—does it include its drop? They do not mention something we think might be important— how much water flows in it—but this depends on when, where, and how it is measured.

Facts and theories in Science are rivers. Their sources are contestable matters of profound triviality. They do not always say what we think they should say. Their meanings and importance change, sometimes dramatically. Facts are for textbooks, and textbooks are often for dustbins as soon as they are written.

Although we may recognize the absurdity in thinking of a river as having *a* source, we do not abandon this idea, for the narrative serves a purpose. "Facts" don't have to be true to be useful; so long as a speaker and a listener are willing to accept a shared premise, they can communicate. Indeed, they can communicate only to the extent that they are willing to accept shared premises. Accepting a shared premise does not mean accepting it as "true" in any absolute sense; rather it involves us in accepting that some statements are conveniently memorable, and that they are demonstrably false only in ways that don't seem to matter. We "accept" such statements for their utility, and for their narrative potency.

Utility is easy enough to understand. It might be useful to know that if we are on the left bank of a river, we are in one country, but if we are on the right bank we are in a different country. If we are Vikings in a longboat and wish to sack Tours, it might be good to know that we can reach there from Saint-Nazaire.

Narrative potency touches on how we remember things, how we make sense of the world. Man is a story-telling animal, and the stories we tell and remember are stories with simple morals— that if one thing happens then another follows, and another, and another. They are stories that construct a causal chain. We like to think that by our reason we can make sense of the world, and that this involves exposing the causal chains by which things happen. Science, it seemed, was the vehicle by which we constructed such an understanding.

In what follows, we explore how scientific facts are constructed, how they are used to build stories, and how those stories spread.

1 The Norms of Science, and Its Structure

Scientists are no different from anyone else. We are passionate human beings, enmeshed in a web of personal and social circumstances. Our field does recognize canons of procedure designed to give nature the long shot of asserting herself in the face of such biases, but unless scientists understand their hopes and engage in vigorous self-scrutiny, they will not be able to sort out unacknowledged preference from nature's weak and imperfect message.

—Stephen Jay Gould[1]

If you ask a scientist what it is that scientists do, she might answer that scientists are engaged in extending our knowledge and understanding of the world. That scientists design and conduct experiments, analyze the outcomes, and draw conclusions. That these conclusions lead to predictions that can be tested. She will probably present herself as in pursuit of objective truth through the application of a rational scientific method. But she will know that things are a little more, well, complicated.

The outcomes of experiments are what scientists speak of as "evidence." Scientists are expected to report their experiments accurately enough for others to repeat them. The failure to reproduce any outcome is potentially a serious matter, one that might call into question (rarely) the integrity of the scientist, (occasionally) her competence, or (commonly) her experimental methodology. Thus, evidence—the "facts" as reported—is not challenged lightly. But "facts"—the observable and reproducible outcomes of an experiment—gain meaning only in the process of interpretation. The conclusions that scientists come to depend on how they interpret that evidence.

Yet the same evidence can yield different conclusions depending on context—including on other evidence that comes to light. For example, levels of the hormone leptin, whose discovery was announced in 1994, are high in obese individuals. Given no other information, this might be interpreted as suggesting that a high level of leptin is a risk factor for obesity. But high levels of leptin are a consequence of obesity: leptin is produced in proportion to the amount of fat in the body. Knowing this, the observed association might be interpreted as implying that leptin has no role in weight gain. Again, this would be wrong— leptin is a potent *inhibitor* of appetite. Knowing this, the appropriate conclusion might be that obesity develops despite high levels of leptin.[2]

The conclusions that scientists come to become accepted through a process of persuasion; scientists are advocates of their chosen interpretation, and through that advocacy they employ rhetoric as well as reason. What we think of as knowledge and understanding are the products of communication: ideas gain influence not from their merit alone, but also from their explanatory power, a power that reflects in part their predictive potency, but also their elegance and simplicity. Some theories become influential despite being notoriously incomprehensible—but, generally, the ideas that gain most traction are those that provide simple explanations of complex phenomena.

However, what is simple to one scientist might be incomprehensible even to a scientist in a closely adjacent field. Scientists disseminate their findings through papers and talks at conferences; in this, they speak to just a few colleagues with shared interests and shared knowledge. These constitute a community that determines how to weigh the strength of evidence, the plausibility of any proposed interpretation, and the importance of the ideas. These communities sometimes align with recognized disciplines, sometimes with a particular scientific society, but often they have no visible structure. Typically, they are self-organizing and self-policing, and they are in constant flux. National boundaries don't mean much when the fifty or so people in the world who might be interested in your work, who know its potential flaws and can appreciate its implications, live in ten different countries, and often not the countries they grew up in. Science doesn't always fit into the disciplines that we are familiar with. At the frontiers of science, where researchers are focused on specific fundamental questions, science is often *undisciplined* in the sense that "disciplines" such as physiology, pharmacology, and biochemistry hold little sway.

These small communities, focused on very particular topics, review each other's papers and grants, organize conferences and workshops, exchange information and research materials, and provide training and career opportunities for students. Most scientists belong to several partially overlapping communities, and ideas spread between them. Ideas can transcend their original application by providing analogies that inspire new explanations, and sometimes have an impact on an area of science very different from that in which they first arose. But new ideas, wherever they might eventually spread, usually emerge within a very small community that, by the scrutiny that it exerts on its members, gives some collective authority to those ideas.[3]

Merton's Norms

The comfortable view of Science is that it is a disinterested activity of gathering objective and unbiased observations, which, by the selfless cooperation of transnational networks of scientists, lead us ever closer to objective truth. In 1942, the sociologist Robert Merton described its ethos as being characterized by four "norms":[4]

Communism: scientists share the fruits of their labors and collaborate; they share methodologies, research materials, and data; and they publish their findings as soon as they can.

Universalism: the validity of scientific outputs is judged not by the authority of the scientists that produced them, but by objective criteria set by the larger community of scientists.

Disinterestedness: scientists act for the benefit of a common scientific enterprise, not for personal gain.

Organized skepticism: claims are exposed to critical scrutiny before being accepted, and open criticism is encouraged.

These norms, according to Merton, were promoted by the "institutions of Science" (academic journals, grant-funding organizations, universities and research institutes) and by the professional societies to which scientists belong. They imposed moral imperatives not only on scientists but also on the institutions. In 1910, Henry Pritchett, the president of the Carnegie Foundation, expressed this forcefully: "A university has a mission greater than the formation of a large student body or the attainment of institutional completeness, namely the duty of loyalty to the standards of common honesty, intellectual sincerity, of scientific accuracy."[5]

However, the institutions have changed. In England, in 1950, just 3.4 percent of students who left secondary school entered higher education; 17,300 students graduated with a first degree; and 2,400 with a higher degree. By 1970, 8.4 percent of students were entering higher education, and over the next thirty years this increased to 33 percent. In 2010, UK universities awarded first degrees to 331,000 students, and 182,600 were awarded higher degrees.[6] This expansion entailed a huge increase in the numbers and size of universities, with profound consequences for their management and culture. Universities are now in the business of attracting income. Their academics are the means for gathering income first, and the point of gathering that income is only a second consideration, often a distant second.

The University of Edinburgh, where we work, is the sixth oldest in the English-speaking world. In the 1970s, scientists here could expect their research to be funded from the university's coffers. Today, scientists at Edinburgh must fund their research by grants from external sources, and, for most, their continuing ability to do so is what assures their continuing employment. Edinburgh is a research-intensive university, one of the top five in the United Kingdom: it is very successful in winning funding from research councils, private industry, charities, and international funders. But while it once could control its research through departmental chairs, that control has been surrendered to the market forces of a complex funding environment. Now, universities measure the value of grants more by the money they bring than by the results that follow; they value the image that they present more than the reality; and they continually reorganize to project the image they think is required.

Institutions in Flux

Over the same period, the culture and role of societies have changed. The Physiological Society in the United Kingdom was founded in 1876 in response to pressure from antivivisectionists to control the use of animals in experiments: its initial purpose was to speak to the government on this matter with a unified and authoritative voice.[7] The nineteen founding members included the most eminent professors of physiology from the Universities of Oxford, Cambridge, Edinburgh, and London. They also included some of the "aristocracy" of academia: the anatomist Thomas Huxley ("Darwin's bulldog"); Charles Darwin's son, the botanist Francis Darwin; the polymath Francis Galton; and the philosopher George Henry Lewes, partner of the novelist George Eliot. Charles Darwin was one of the two original honorary members.

The early society was a Victorian "gentleman's club." Meetings, mostly at fashionable London restaurants such as the Cafe Monico on Shaftesbury Avenue, began with a dinner followed by a discussion of business. One role that the society gave itself was that of recommending to the grants committee of the Royal Society "such researches in physiology as it might be thought desirable."[8] In 1878, its members founded the *Journal of Physiology*, the first English language journal devoted to this discipline. In 1880, they organized their first scientific meeting. This meeting, to which several foreign guests were invited, included demonstrations of experiments: "they were to be private and confidential [. . .] with the view of eliciting remarks and criticism of work in progress."[9]

A hundred years later, in 1980, the society was organizing six to ten scientific meetings each year, hosted by universities throughout Britain and Ireland. At any meeting, a member could elect to present, in a ten-minute talk, the results of their recent experiments, for the approval of the society members, or they could introduce a nonmember to deliver such a talk. Approval was not a formality: after each talk, there were questions and discussion, followed by a vote on whether the abstract of the talk should be published in the *Journal*. It was a process that exposed scientists to often acerbic criticism, and it led to lively debates and wounded pride.

A supplicant wishing to join the society in 1980 had to have published some papers in physiology, preferably in the *Journal of Physiology*. They had to have attended meetings of the society and presented research findings at them. They had to hold a tenured appointment at a university or research institute, or show other evidence of a commitment to physiology. They had to solicit the support of members from diverse institutions to act as their nominators. If the application was deemed credible by a committee, their name would appear on a ballot for election by a vote of all members—with votes against counting five times each vote for. If elected, they could take their place amongst the voting members who judged the worth of the research presented at meetings.

The society was run by officers elected by the members. These roles were onerous, but were understood to be markers of esteem. Universities supported the society by providing venues for meetings free of charge, encouraging staff to become involved in societies, and financially supporting their attendance at meetings.

The Physiological Society was just one of many that exerted academic authority over their disciplines, but this had started to change by the end of the 1990s. With the growth of science and its increasing internationalization, scientists became less secure and more mobile, expecting to spend years on short-term contracts in different institutions before finding settled employment, often moving from one area of research to another or from academia into private industry.

With technological advances came a diversification of experimental techniques, each requiring specialized expertise. In the 1970s, most members attending a society meeting would have been familiar enough with the methods of any experiments to contribute to a discussion of the outcomes, but this ceased to hold true. Talks became advertisements for notable findings, rather than invitations for advice, and the tradition of open criticism vanished, as a perceived anachronistic exercise of machismo.

Accordingly, societies that once restricted membership to those who had achieved a certain recognition became inclusive. The Physiological Society, like many others, dropped its membership requirements and encouraged anyone who wished to join. It appointed professional executives to run the society, reduced the number of meetings and increased their size. Universities ceased to subsidize them and began to look askance at staff who gave much time to supporting the activities of a society.

This decline of elitism—of hierarchically structured authority—had inevitable implications for how knowledge is made and spread, but these went unnoticed.

The Norms of Science, and Conflicting Expectations

Each of Merton's norms is often breached.

Communism is in conflict with competition. Scientists share the fruits of their labors, but sometimes forget inconvenient results. They collaborate when it is in their interests, but not when they are in a race for priority. They share research materials—after publication, after the races have been won. They publish details of their methods, but sometimes withhold enough to retain an advantage. They share only some data, often only with trusted colleagues, and usually withhold unpublished data. Sometimes they publish quickly, but often wait until they have a complete story that they can "sell" to a major journal.

Universalism asserts that the scientific outputs are judged by objective criteria. But are there any objective criteria? Every scientist thinks that their own findings and ideas are important.

Disinterestedness asserts that scientists act for a common scientific enterprise. Yet they pursue personal prizes and accolades. They are not paragons; they cannot always lay aside their various biases. They are not immune to the hope of financial profit, nor are sexism and racism strangers to their ranks.

Organized skepticism requires that claims be critically scrutinized. Scientists need collaborators, as well as friends on grant committees and editorial boards. Does this temper how their skepticism is expressed? Do they ever change their minds?

There is nothing new about this. The norms have been breached in such ways throughout the history of Science. They have been respected as ideals, but have not served as a rigid rulebook.

Scientists are people, people with preconceptions and biases, striving to build their reputations and advance in their careers, beset by a host of external pressures. Commercial funders want products that will sell. Charitable funders have single-minded priorities that are sometimes distant from the actual problems. Politicians want quick responses to public concerns at as little cost as possible, and ideally ones that attract good publicity. Governmental funders might aim to fund the best science, but must account to politicians anxious to respond to public perceptions. The media is hungry for headlines: pressure groups and lobbyists agitate against stem-cell research, nanotechnology, biotechnology, and for more of the available funding to be spent on their particular concerns. Peer review can be compromised by jealousy and competitive interests and by subjective interpretations of what is important.

Yet most scientists do uphold Merton's norms—in principle, if not always in practice. The pressures on scientists are not unidirectional; they often conflict with one another, and scientists are not passive, any more than any person is when faced by a similar array of pressures in their lives. Scientists, like all people, resist such pressures. This resistance might sometimes be fragile; sometimes, scientists wittingly or unwittingly, in serving the interests of funders, compromise their scientific message. But many scientists reject funding if they suspect that accepting it might compromise their work. Many others refuse any personal remuneration from external sources, and apply all funds received to research.

The Construction of Scientific Understanding

Many scientists strive to explain to the public why they believe that what they do is important. Some become partisan advocates for particular policies. Some try to "educate" the public, perhaps by making the case for why the "obvious" solution to a problem won't work. But all of these are different from trying to explain what scientists actually do. Few really attempt *that*. Science at "the cutting edge," where most scientists would claim to be, is a murky place, beset with doubts, disappointments, anomalies and contradictions. Progress comes often seemingly invisibly or from unexpected directions. Arthur Clough's poem, "Say not the Struggle nought Availeth," captures this sense:[10]

For while the tired waves, vainly breaking
Seem here no painful inch to gain,
Far back through creeks and inlets making,
Comes silent, flooding in, the main.

But few of these frustrations penetrate the published literature. In what they write, scientists assume an air of confidence and certainty about their own work. At the same time, in carefully chosen phrases, they undermine the authority of rivals. Strong narratives don't leave room for circumspection and humility: the meek do not inherit the accolades that come from impact, and when scientists acknowledge the limitations of their own work, this has often forced been upon them by editors and reviewers.

Some think that Science approaches an absolute truth about what the world is like; that while individual scientific finding may be unreliable, the "main" carries some semblance of that truth. By this view, a particular scientist might be known for a landmark advance, but, in the long run, such advances are inevitable—if one scientist hadn't reached a particular landmark, it would have been only a matter of time before another did. Scientific understanding is there to be uncovered and is fixed by the nature of the world.

An alternative view is that scientists construct our understanding, and that this understanding is constrained by the ways available to us to view the world and by the limitations of our brains to process and interpret evidence.

By this view, our understanding is path-dependent. How we understand the world depends on the influence of others who have gone before, and hence on *their* biases, prejudices, and preconceptions. This understanding can be augmented by fresh evidence, but to find evidence we must look for it—and what we will seek depends on what we think, at a given time, is most important to know, and on what methods are available to search. Often, a methodological advance will change the course of Science, as some neglected problems become important simply by becoming tractable. Our understanding of the world changed with the light microscope and again with the electron microscope; these, in changing what we can see, changed what questions were important.

We see the world "through a glass darkly," through a lens clouded by our prejudices and preconceptions. Where scientists look with their imperfect lens is guided by pressures within and beyond the scientific community, and by opportunity. A policeman, so it is told, came one night upon a man looking for his keys by the light of a streetlamp, and he asked the man where he had dropped them. The man pointed at the darkness. "So why are you looking here?"—"Because this is where the light is."

The Structure of Science

Science is complex. Scientists create and sustain a flexible structure that enables them to pursue their research, but this structure imposes restrictions on them. It defines, for any particular question at any particular time, how they design their experiments, how they validate their methods, how they interpret the outcomes, how they present their work to others, how ideas are understood, and how ideas spread. These definitions determine who will be successful and whose work will languish in obscurity.

Philosophers have focused on the logical content of knowledge, and in so doing they have created myths about the rationality of Science. But many published findings are false. Fraud is rare, but questionable practices barely short of fraud are not.[11] Fundamental statistical weaknesses are endemic; conflicts of interest distort published outcomes; and common journal practices and citation practices subvert the integrity of the literature. These findings can only be understood from a social perspective, but their implications for the validity of knowledge require epistemological and scientific insight. For the challenge is not to understand how rational and unbiased minds harness the accumulated wisdom of their forbears and use critical reasoning to design experiments that progress our understanding. Rather, it is to understand how progress occurs despite the failings of reasoning, and despite the myriad of pressures on scientists.

Perhaps we need also to consider whether the same forces that give rise to the failings of Science also give rise to its success. It is claimed by many that unfettered capitalism, a system that gives rise to extreme inequalities and inequities, environmental degradation and human exploitation, nevertheless advances the general good more effectively than centrally controlled economic systems. It is claimed that it does so by fostering a Darwinian struggle that ensures ever-increasing productivity and technological innovation, resulting in escalating societal wealth. The merits of this case in the domain of economics are not for us to consider here, but we might at least consider its merits in the domain of Science.

Generally, we must consider how the structure of Science affects the nature of our understanding. The nature of scientific communities and institutions and the mechanisms by which evidence is evaluated and by which ideas are disseminated are all in flux, and there is increasing concern about the reliability of science. In the following chapters, we ask how science actually works, and we begin with the two philosophical views of how science progresses that are best known by scientists.

2 Popper and Kuhn, and Their Conceptions of What Science Is

> I cannot give any scientist of any age better advice than this: the intensity of the conviction that a hypothesis is true has no bearing on whether it is true or not. The importance of the strength of your conviction is only to provide a proportionately strong incentive to find out if the hypothesis will stand up to critical evaluation.
>
> —Peter Medawar[1]

We expect that, when governments make decisions that affect how we live, they should be based on scientific evidence. Such evidence is available through the scientific literature, but it requires interpretation and evaluation. For these things, we cannot take at face value what authors say of their own work, for scientists are partisan advocates of the validity and importance of their work.

Evidence seldom "speaks for itself"—it must be spoken for. It is spoken for by people, and people whose passion and clarity of thought and expression make their message persuasive, who speak to a receptive audience seeded with friends and disciples, are those who are most influential. We cannot understand Science except as a social activity. This is not to diminish it, but to accept that scientists are in the business of influencing people.

Man is a story-telling animal—we remember stories better than facts, and often misremember facts to fit a favored story. The vagaries of our memories have been extensively catalogued: we attend more to facts that conform to our expectations and prejudices, recall them selectively and imperfectly, and, when forced to acknowledge inconvenient facts, we strive to interpret them in a way that fits our preconceptions.[2]

For each of us, our understanding of the world is based on evidence: it may be evidence in the Bible, or evidence of our experience, or evidence of what we have heard from others or read in newspapers. With this evidence we construct stories that help us to understand the world in terms of causality—why things happen the way they do. But we also use evidence to propagate those stories, to influence others to behave in a particular way, to vote for a particular candidate, to buy a particular product, or to treat certain people differently. In these cases, we use evidence selectively; we choose to tell things that support whatever story we are telling—we cherry-pick the evidence, and sometimes we twist it.

Scientists also build stories upon evidence. Their evidence includes "harder" evidence, collected by rule-governed processes and subject to inspection and confirmation. Yet there is always some subjectivity, some bias. Scientists collect the evidence that they look for, and they look for what they expect to find and need to find to support their stories, the stories that they call theories. Theories determine what evidence they look for, how to look for it, and how to interpret it, so there is a circularity.

As scientists elaborate their theories and test them, all the while they are battling to persuade others of the validity and importance of their stories. This involves disparaging the validity and importance of rival stories, and in these battles they also use evidence as ammunition.

Scientists have a certain responsibility when it comes to how they use evidence. They are trusted, and their advice is given respect. We do not want to argue that that trust is misplaced. To know how something works, it seems wise to listen to people who have spent a long while studying exactly that, and who have a lot invested in being, most of the time, right. But scientists, because of that responsibility, must, in Gould's words, engage in "vigorous self-scrutiny."[3] For they are not immune to failings that are common to all people. If the progress of science depends on how scientists gather evidence and on how they use it, then these failings are things we must play close attention to.

For Science faces profound challenges. Even the top rank of the scientific literature is not an archive of pristine truth; many of the most influential papers and journals contain irreproducible findings.

Karl Popper: Good Science and Poor Science

We have to begin by looking at how scientists construct their stories. Philosophers have long tried to understand what, if anything, makes Science special. The most influential of these, at least as reckoned by his influence on scientists, was Karl Popper (1902–1994), with the publication of *The Logic of Scientific Discovery*.[4]

Popper began by trying to understand what distinguished science from pseudoscience. Growing up in prewar Germany, in a community polarized by Marxism and fascism, he had seen that the ideologues were impervious to experience—everything they experienced seemed to reinforce their particular worldview. For Popper, whatever science was, it wasn't that.[5]

Before Popper, the dominant movement in philosophy was *logical positivism*—the view that the only meaningful statements were those that could be verified empirically. This portrayed science as a progressive activity, building from facts, drawing inferences from them, and confirming those inferences by further empirical studies. Popper destroyed this edifice.

His attack was direct and incontrovertible: "No matter how many instances of white swans we may have observed, this does not justify the conclusion that all swans are white."[6] With this, Popper dismissed the idea that the truth of any nontrivial hypothesis can be established through any amount of supporting evidence. The criticism was a rearticulation of David Hume's Problem of Induction, posed over two hundred years years before.[7] But Popper proposed a solution. He argued that the only logically valid path to certain knowledge is *refutation*—a single instance of a black swan would be enough to disprove the hypothesis that all swans are white. Hence, scientists should not strive to confirm hypotheses, but to refute them.

Popper concluded that scientific statements differ from pseudoscientific statements in being capable of refutation: "In so far as a scientific statement speaks about reality, it must be falsifiable: and in so far as it is not falsifiable, it does not speak about reality."[8] By this view, scientists are professional skeptics, their knowledge forever open to doubt. This, for Popper, distinguished scientists from those whose beliefs are held on trust, faith, ideology, or superstition, who find confirmation in everything they see, and so can make no predictions by which their beliefs might be refuted.

But if little worthwhile understanding can be gleaned from accumulating "facts" and generalizing from them, how does science progress? For Popper, the answer lay in the imagination of scientists—their ability to devise theories that leapt beyond any direct observables. Theories were built on provisionally accepted "facts" that comprised observations as interpreted through those theories. Science in this sense was a rickety, self-supporting structure:

> Science does not rest upon solid bedrock. The bold structure of its theories rises, as it were, above a swamp. It is like a building erected on piles. The piles are driven down from above into the swamp, but not down to any natural or "given" base; and if we stop driving the piles deeper, it is not because we have reached firm ground. We simply stop when we are satisfied that the piles are firm enough to carry the structure, at least for the time being.[9]

Popper gave scientists a reason for celebrating the creative role of the scientist,[10] and ways not just of telling between science and pseudoscience, but also of telling between good science and poor science. Poor science sought empirical support for that which we had reason to expect to be true; good science challenged our preconceptions. Good science led to bold predictions by which theories could be refuted, and the theories that survived rigorous testing thereby earned their place in our understanding.

Thomas Kuhn and the History of Science

How far scientists actually followed this path was the subject of *The Structure of Scientific Revolutions* by Thomas Kuhn.[11] Kuhn thought that the history of science was more than an archive of anecdotes, but contained evidence from which we might understand why Science is so successful.

Kuhn sought to give an account of scientific history that respected context—one that didn't judge discarded scientific theories on the basis of what we believe today. For example, Aristotelian physics has been caricatured as either not science at all, or to the extent in which it is science, as blatantly wrong.

Commonly, Aristotle's "errors" are represented as arising because he tried to understand the world through reason alone, neglecting empirical evidence. This is just wrong.

Aristotle was one of the earliest thinkers to have proposed that "natural laws" governed the world. For his mentor, Plato, epistemology—the understanding of the nature of knowledge and the nature of justified belief—began with knowledge of universal "forms" (or ideas) which led to knowledge of particular instances by logical deduction. But Aristotle was an empiricist, and his epistemology was based on the study of things in the world. Such studies, for Aristotle, led to knowledge of universal principles through a process that we might now recognize as "inductive reasoning." Aristotle's work on biology and geology comprises a substantial body of systematic observations. In his *History of Animals*, he recorded his observations of reproduction in birds, of the formation of eggs, and of the embryos within them. He recognized that developing organisms go through a series of stages before acquiring their final form and he proposed "principles of generation" to account for this.[12]

But Aristotle's laws of motion seem now to make little sense. His definition of motion (change of any kind) is commonly translated into English as the "actuality of a potentiality"—a translation that seems to have no meaning whatsoever.[13] However, in the thirteenth century, St. Thomas Aquinas (1225–1274) decided that Aristotle meant what he wrote and that what he wrote is worth trying to understand, and he produced a clear and coherent commentary on Aristotle's notions.[14] Aristotle understood that plants and animals had within them a principle of motion—an innate impulse to change. The embryo is a chicken; it is not defined only by the characteristics that it currently exhibits, but also by the potential to become something else. *Motion*, as Aristotle used the term, is the mode in which that potential is achieved. He extended this to all "natural things"—the movements of the stars and planets he ascribed to innate impulses, and the elements (earth, fire, air, and water). All seemed to have innate principles of motion by which each could be distinguished from the others. Fire moves upward, because it is natural for it to be there; by contrast, artificial things like a bed or clothing, have no innate principle of motion, except insofar as they are made of natural things. A knife contains a principle of downward motion, not because it is a knife, but insofar as it is iron. Not every motion involves an innate principle—the principle of motion could come from an external agent. In things that move naturally, like fire and the stars, there was an active principle, but other things had a passive principle—matter—and for these, motion required a mover. From his observations, Aristotle inferred a relationship between the size and weight of an object and the speed at which it moves either when propelled by an external force or when falling to ground by its innate tendency to move down—relationships that we now call his laws of motion.

Kuhn showed how Aristotle's laws made sense in the context of what Aristotle meant and observed. In contemplating a falling stone, Aristotle saw a change of state, not a process that could be characterized instant by instant. The speed at which the stone fell was simply a function of distance travelled and time taken— what we would now understand as average speed.

Kuhn's analysis is supported by the physicist Carlo Rovelli, who argued that, so far from being a crude, intuitive understanding "Aristotle's physics is counterintuitive, based on observation, and correct (in its domain of validity) in the same sense in which Newtonian physics is correct (in its domain)."[15] In Newtonian physics, a stone falling in a vacuum (the domain of Newtonian physics) is subject to constant acceleration, but a stone falling through air or water (the domain of Aristotelian physics) rapidly achieves a constant velocity, as observed by Aristotle.

From such case studies, Kuhn concluded that progress in science is not a simple process of replacing one theory with another that better fits new evidence. Theories make sense only when understood as embedded within a much larger domain of understanding. It is difficult to compare theories that are talking about different things, and even when they use the same terms, they attach different meanings to them. Such problems, Kuhn argued, do not apply only to theories developed at different times, but also to competing, contemporaneous theories.

Rather than asking how Science progresses by the traditional philosophical approach of understanding progress as the consequence of refinements in the scientific method, Kuhn adopted a sociohistorical approach. He held that it is incontrovertible that Science has made remarkable achievements. Science is the product of what scientists do, so to understand how it has progressed we must study how they acquire and use evidence, and in what sense they extend our knowledge.

We know the fallibilities of individual scientists, the false trails, blind alleys, and pitfalls that inevitably accompany any journey into the unknown. These do not concern us here. You cannot discover a safe path without defining the verges and exploring the undergrowth for hazards, nor can you find the best path without surveying others. Rather what concerns us is whether some of those false trails, blind alleys, and pitfalls are of scientists' own making—the avoidable traps.

If we do this, then we must also consider it possible that, however absurd, irrational, and inefficient scientists might seem to be, some of those absurdities, irrationalities, and inefficiencies might nevertheless contribute to the progress of science. This is not necessarily so—science might progress despite egregious imperfections—but Kuhn recognized that the question must be asked.

Kuhn's Analysis of Science

Kuhn argued that, most of the time, scientists work within a *paradigm*. The concept of a paradigm was elusive; conveniently so, as Kuhn could use it for different things at different times and in different contexts. But a paradigm is generally taken to mean a framework that emerges from the activities of a community of scientists. It includes a body of knowledge and theories; it defines what questions are important and what methods are appropriate to answer them; and it defines the standards by which scientists and their results are judged—how evidence is to be weighed.[16] Kuhn concluded that "normal" science was less about building and testing theories than about "puzzle solving": small, incremental advances that elaborate theories and extend their predictive potency.

In Kuhn's reading of history, paradigms are not, generally, the subject of assault and are not easily discarded when contradictory evidence emerges. Only when substantial anomalies accumulate does any real challenge arise. Then, the challenge arises less because scientists acknowledge the failings of a paradigm than because an alternative paradigm arises, one that succeeds in attracting new followers.[17]

A "scientific revolution" can be recognized once it becomes apparent that a new paradigm has replaced an older one, but this is not to say that the scientists working in the older paradigm are converted. Kuhn argued that scientists do not readily change their minds or ways of working, and a paradigm finally dies only when its followers have abandoned science altogether.[18] He quoted Max Planck: "A new scientific truth does not triumph by convincing its opponents and making them see the light, but rather because its opponents eventually die, and a new generation grow up that is familiar with it."[19]

Kuhn argued that this conservatism is not irrational resistance to change. There is always an extensive investment in any paradigm—intellectual investment and investment in equipment and other resources—and this investment is not abandoned lightly. Moreover, it is seldom clear that a new paradigm is better than the old. Often, at least at first, the converse is true: a new paradigm might explain some things better, but in other respects, while it is still raw and provisional, it might be able to explain less than the old. The choice between paradigms cannot be resolved simply by weighing the evidence—other factors are important.

Generally, the proponents of competing paradigms use different experimental techniques, weigh some sorts of evidence more highly and other sorts less highly, and define different questions as important. According to Kuhn, competing paradigms are "incommensurable": the worth of one cannot be assessed by the values of the other.

Problems and Puzzles

The *Logic* and the *Structure* were masterpieces of persuasion. Each sustains a fluent and coherent narrative punctuated by irresistibly quotable passages. The *Logic* vividly expresses passion, and the core argument is expressed in memorably simple terms.

But "all swans are white" is a very simplified version of a hypothesis, one that Popper introduced as a thought experiment. It was useful in exposing a fundamental weakness of common reasoning. But scientific hypotheses are not mere generalizations. They take many forms, and often include statements that assert a causal relationship of the form "*A* happened because *B* happened." Generally, such statements are not the same as "All instances of *A* are followed by instances of *B*." Smoking causes lung cancer, but most smokers will not get lung cancer and some people will who have never smoked. Whether smoking results in lung cancer in any individual will depend on many other things, including on chance, genetic predisposition, and other factors that we do not understand.

Conversely, even if all instances of *A* are always followed by *B* this would not establish any causal connection. Birds sing before the sun rises, but the sun does not rise because they sing. To assert a causal relationship require more than evidence of association. We need to understand the mechanism by which one thing causes the other—evidence that tobacco smoke contains carcinogenic substances. And we need evidence that we can break the association by interventions—that stopping smoking will reduce the incidence of lung cancer. Constructing a strong hypothesis requires assembling different types of evidence: evidence of association from observational studies, evidence from intervention studies, and evidence of mechanism that requires interpolation of more "fundamental" theories. Finally, this evidence must be compiled to form a persuasive narrative.

Kuhn weaved his own narrative through familiar highways of the history of Science, through Aristotle, Copernicus, Galileo, Newton, Dalton, and Einstein, names familiar from busts and hagiography. His is a view of Science formed through inductive reasoning—the form disparaged by Popper. It assumes that the present is like the past, and is strangely at odds with the view that Science is a social product, its nature determined by present circumstances. Like Popper, Kuhn used rhetoric in a very effective way, in a roller coaster of narrative conveyed by concepts like *paradigm* and *revolution* whose rhetorical strength lay not in their precision but in their very imprecision—in their ability to be read in different ways at different points in his narrative, ways that suited the particular reader at each point.

Nevertheless, a generation of scientists was inspired by Kuhn and Popper. From Popper, they embraced the vision of scientist as a creative enterprise, the principle of skepticism, the merits of hypothesis-driven science, and the search for "killer experiments" that would refute a theory. From Kuhn, they were inspired by the notion that it is the young that spearhead innovations, by the idea of perhaps thus contributing to a new scientific revolution. From both, they were taken by the notion that older scientists need not be unduly respected.

Popper had identified a mode of argument that, in Science, is particularly persuasive. If a scientist can propose a clear and bold hypothesis, design experiments to test it rigorously, and obtain clear outcomes whose interpretation is hard to contest, then she will have the material to construct a narrative that is likely to be persuasive. Of course, whether she actually succeeds in persuading others also depends on being able to convey that narrative effectively. Both Popper and Kuhn were masters at this.

3   *Laboratory Life*: Bruno Latour and Rhetoric in Science

This appearance of success cannot in the least be regarded as a sign of truth and correspondence with nature. Quite the contrary, suspicion arises that the absence of major difficulties is a result of the decrease of empirical content brought about by the elimination of alternatives, and of facts that can be discovered with their help. In other words, the suspicion arises that this alleged success is due to the fact that the theory, when extended beyond its starting point, was turned into rigid ideology. Such ideology is "successful" not because it agrees so well with the facts; it is successful because no facts have been specified that could constitute a test, and because some such facts have been removed. Its "success" is entirely man-made.

—Paul Feyerabend[1]

Whether Popper's valedictions amounted to either a credible account of Science as practiced or a feasible account of how it might be practiced have exercised many philosophers. Paul Feyerabend, a student of Popper who became a dissident, argued in *Against Method* that Science has no governing principles: it encompasses a great variety of processes, and any particular science is a disparate collection of subjects, each containing contradictory tendencies.[2] Advances often occurred because some scientists either decided not to be bound by conventional rules, or because they unwittingly broke them. The history of Science does not just consist of evidence and conclusions, but is filled with ideas, disputes about interpretations, and mistakes. No interesting theory ever agrees with all the facts. In any case, all "facts" either contain assumptions, or are used in ways that assert assumptions—our view of the world is shaped by assumptions that are not themselves exposed to criticism.

Accordingly, some concluded that scientific "facts" could only be understood by considering the context in which they are used and the history of their construction, and by identifying the embedded assumptions and preconceptions. This suggested that sociologists must invade laboratories to see how scientists worked and how they constructed scientific facts.

One such foray became *Laboratory Life*, written by a young French academic, Bruno Latour, in collaboration with an English sociologist, Steve Woolgar.[3] Latour had trained in philosophy, had studied for a PhD in theology, and had then taken part in an anthropological study in the Ivory Coast. Between 1975 and 1977, he spent two years in the laboratory of Roger Guillemin at the Salk Institute in California, a laboratory then racing to identify pituitary hormone-releasing factors. These factors had been postulated twenty years earlier, and it was believed that their identification would lead to important medical applications. Guillemin's success in this would win him a share of the 1977 Nobel Prize in Physiology or Medicine.

Latour made little attempt to understand the work of Guillemin's laboratory on the scientists' own terms. He was not a biologist and did not try to master the technical issues; while his study was of "the routine work carried out in one particular laboratory," he wanted a generalizable understanding of what scientists do. He thus set out to study the "anthropology of science," conceiving his project as "analogous to that of an intrepid explorer of the Ivory Coast, who, having studied the belief system or material production of 'savage minds' by living with tribesmen, sharing their hardships, and almost becoming one of them, eventually returns with a body of observations."[4]

This style of research was novel, and so was the literary style of *Laboratory Life*. It records random snippets of conversations, these "observables" are interpreted as evidence to adduce fact-like statements that purport to express generalities about Science. The book is entertaining—the humor comes from the incongruities: the notion that anecdotal snippets of text and conversation can produce anything meaningful except in a sophisticated mind primed with preconceptions and embedded assumptions; and the notion that such a mind is present in that of a naive fly on the laboratory wall.

Laboratory Life also expresses surprising insights. Despite its manifest failings, eyewitness testimony has particular agency. We don't like to believe that evidence given in good faith by witnesses is contaminated by their prejudices or that their senses are unreliable, because to doubt them is to doubt ourselves. Although, apparently, based on scant and partial observations made in a singular context by a *faux naïf* observer, *Laboratory Life* reads mundanely and meticulously true to scientific life.

Latour saw that the output of the scientists consisted mainly of published papers, the most important of which were apparently those that were read by relatively few scientists, who in turn wrote papers that were read by the scientists whom Latour was observing. The value of each paper appeared to be reflected in how often it was cited and in how it was cited.

He noted that each paper included many statements expressed in ways that appeared to qualify their credibility. These ranged

from type 1 statements, which represented conjectures or speculations, to type 5 statements, which asserted fact-like entities.
The scientists pored over these, paying closest attention to those
that they felt they were identified with, to check that they had
been cited (and cited correctly) and to analyze how each differed
from similar, previously published statements. The activity in
the laboratory (including what we would call the collection of
evidence) was directed towards transforming statements from
one type to another, with the aim of generating credit for as
many statements of type 4 as possible (because type 5 statements
represent "facts" that could be taken for granted, they were generally given without specific attribution).

> The anthropologist feels vindicated in having retained his anthropo-
> logical perspective in the face of the beguiling charms of his infor-
> mants: they claimed merely to be scientists discovering facts; he
> doggedly argued that they were writers and readers in the business of
> being convinced and convincing others.[5]

The shock delivered to the reader of *Laboratory Life* is that
of "why didn't I see that before?" One might wonder whether
the authors couldn't have made their case without ever setting
foot in the laboratory. Perhaps they could have, but their work
wouldn't have the same persuasive power without the rhetorical
devices that it assembled. Their insights are not deduced from
the evidence that they assemble, but are illuminated by it. This
same sense of shock makes the narratives of Popper, Kuhn, and
Latour so powerful. For each of these authors, "evidence," as
they use it in their own books, is essentially a rhetorical device.

Returning to Merton's norms, we now might see that these
are not moral imperatives. *Communism*, that scientists freely
share the fruits of their labors and publish their findings as soon
as possible, is a pretentious expression of a truism. The outputs

of scientists are the papers they write, and it is by these that we recognize them as scientists. *Universalism*, that scientific outputs are judged not by the authority of the scientists but by objective criteria, is merely an account of what modes of persuasion are effective in modifying the status of statements. *Disinterestedness*, that scientists act for a common scientific enterprise, reflects the fact that scientists advance in their careers according to how their statements gain credibility in passage through the scientific literature. *Organized skepticism* describes how scientists, in elevating some statements to a higher type-level, simultaneously derogate others.

By Latour's analysis, the research work of the laboratory consists in generating data that can be transformed to provide "evidence" that is employed in the service of persuasion. To this end, evidence must be credible—open to replication—and some types of evidence are more persuasive than others. Evidence that refutes a particular statement has particular agency—Popperian falsification, however rarely achieved, can relegate a statement of any type to oblivion. There are many types of evidence, and not all claims and hypotheses are amenable to falsification, but if a paper is published that tries to refute a type 5 statement, the consequences for the credibility of the work of other scientists may be serious. Unsurprisingly, such attempts can meet with stiff resistance, as we will see in chapter 5. Nevertheless, the good scientist ("good" in a Popperian sense), is constantly engaged in attempting to refute their own ideas. What happens when they are successful in this is the subject of the next chapter.

4   Is the Scientific Paper a Fraud? The Place of Hypotheses and Their Falsification

In 1960, Peter Medawar was awarded a share of the Nobel Prize in Physiology or Medicine for his discovery of acquired immunological tolerance, a mechanism that was important to understand if organ transplantation was to become viable.[1] That same year he gave a talk titled "Is the Scientific Paper a Fraud?"[2] In it, he did not suggest that scientific papers involve any deliberate misrepresentations but rather that they embody "a travesty of the nature of scientific thought."

The typical paper of that period, as Medawar explained, began with an introduction, which gives an account of previous work and of a key question left unanswered. This was followed by a methods section that describes what experiments were performed, a results section that reports the experimental outcomes, and a discussion that presents an interpretation of this evidence and portrays its importance.

There are good reasons why scientific papers should be formalized. The aim of all good writing is to propel an idea from one mind into another, and, in scientific writing, it is important to do so clearly and concisely. The actual process of discovery is often messy, but the reader need not know about all of the errors and false trails, the failures and frustrations. These "sanitized" accounts may mislead the unwary by hiding the messy reality of scientific life, but they deliver a clear benefit.[3] Separating the sections allows each to be judged on its particular merits. The introduction can be judged by how well it presents the current state of knowledge; the methods by the appropriateness of the experimental design and the completeness of its account; the results by the completeness of the description of the outcomes and their analysis; and the discussion by its persuasiveness.

This structure defines the narrative but distorts the reality. The introduction feigns ignorance about the outcomes of the experiments to be described. The methods section excludes any explanation of why the experiments were designed the way that they were. The results section excludes any interpretation of the outcomes and leaves the sequence of experiments unexplained. All interpretation—the subjective elements of a paper as distinct from the objective details of methodology and outcomes—is relegated to the discussion.

Such narratives, according to Medawar, misrepresent the nature of scientific thought by hiding the role of *ideas*. No experiments are conducted without some idea of what they are expected to show—a hypothesis that both provides the reason for doing the experiment and determines how the results will be interpreted. Such narratives present experiments as though they add predictably and incrementally to knowledge, building a wall of scientific authority brick by brick. They neglect the personal and creative contribution of the scientist, the contributions that decide where that wall will be, what height and color, and what purpose it serves.

In subsequent years, hypotheses figured large in scientific papers. It became usual to use the introduction to state a hypothesis and explain the reasoning behind it, whereas the rest of the paper presented the experiments as a test of it. However, if hypotheses are, as Popper argued they should be, bold ideas, whose refutation (or survival despite attempted falsification) marks the progress of Science—then we should expect to see the scientific literature dominated by narratives of their destruction. It is not. On the contrary, it is a voluminous archive of their confirmation. So, what is going on?

We must consider what scientists are remembered for. Typically, it is not for things they got wrong, but for things they can claim to have gotten right.

Dionysia Theodosis and Her Bold Theory

In 1981, Dionysia Theodosis, a neuroanatomist working in Bordeaux, was studying a small part of the brain. Neuroanatomists study the fine structure of the brain, typically, the details of how particular neurons in the brain are connected to other neurons. The neurons that Theodosis was interested in make oxytocin—a hormone that is essential in lactation—and they release it from their nerve endings in the pituitary gland. Oxytocin causes milk to be let down at the mammary glands when infants suckle, and it is secreted in large pulses. These pulses are the result of intense, synchronized bursts of electrical activity of the oxytocin neurons.

These bursts are only seen in lactation, and so Theodosis speculated that lactation is accompanied by some change in how the neurons interacted with each other. Using electron microscopy, she noticed a change in the glial cells that surround the oxytocin neurons.[4] Glial cells were thought of as "supporting cells" that kept the environment of neurons stable. To many neuroscientists they were merely the "polystyrene packing" of the brain, but to neuroanatomists they were enigmatic. They have long processes that flatten out like sheets of gossamer, sheets that wind around the neurons. These sheets enfold the neurons, nurturing them and protecting each from the bustle of their noisy neighbors. Theodosis found that the glial cells around oxytocin neurons looked different in lactating rats. The processes had retracted: the sheets no longer separated neuron from neuron—oxytocin neurons were now touching each other, and were thus, seemingly, obliged to listen to each other. She proposed that it was this that allowed the neurons to burst in synchrony. This was a bold hypothesis, one to which she devoted much of her career.

Between 1981 and 2006, Theodosis published many papers developing this hypothesis. She showed that the changes first appeared at the end of pregnancy and disappeared at the end of lactation. She showed that the changes could be elicited by oxytocin itself, and that they were promoted by the steroid environment of pregnancy and lactation. She showed that the changes only affected the oxytocin neurons and not adjacent neurons that made a different hormone, vasopressin. She showed that the glial cells could influence the transmission of signals to the neurons that they engulf. From her published output we recognized thirty eight papers as relevant, including studies of how the changes arose and review articles that discussed their functional significance. By 2006, these had amassed 2,143 citations, and eight papers had each been cited more than a hundred times.[5]

But while Theodosis had gathered an array of circumstantial evidence, not all observations fitted her narrative. And, as was pointed out by skeptics, the hypothesis was still to be critically tested. The "killer experiment" was easy to propose—if glial changes were essential for bursting, then blocking them should prevent the bursting, but to do this experiment required finding some way to stop the changes from happening. By 2005, Theodosis had found a way: she showed that an adhesion molecule, polysialic acid, was critically important—eliminating this molecule with an enzyme, endoneuraminidase, prevented the changes in glial cells.[6]

A young scientist, Gwénaëlle Catheline, was given the task of conducting the killer experiment. If Theodosis's hypothesis was right, endoneuraminidase would prevent the suckling-induced bursts. In 2006, the outcome was reported in the journal *Neuroscience*, and the abstract summarized the disappointing news:

> After bilateral endoneuraminidase injections [. . .] neither the frequency nor the amplitude of suckling-induced reflex milk ejections was different from vehicle-treated dams. [. . .] Basal and bursting activity characteristic of oxytocin neurones before each reflex milk ejection was not significantly different from that recorded in [. . .] rats with normal levels of polysialic acid.[7]

Theodosis's bold hypothesis was refuted: "Our results indicate that neuro-glial remodeling [. . .] is not essential to parturition and lactation."[8] After Catheline's paper, interest in the thirty-eight papers that Theodosis had published faded, at least as measured in how often they were cited. Catheline's paper had an impact in redirecting attention to other possible answers to the puzzle of how bursting is orchestrated in oxytocin neurons. It had not cast doubt on any of the results published in those thirty-eight papers—but their interpretation was now uncertain. The narrative that sustained interest in this work was broken, and Catheline's paper has, in the twelve years since its publication, been cited just thirty-one times. This was the end of a story that nobody was going to pursue, and, soon after, Catheline left the field.

Is the Scientific Paper a Fraud?

We measure the quality of scientists by imperfect measures—how often their papers are cited and which journals they are published in. Those rare papers that, like Catheline's, meet the high demands of Popperian science are seldom cited because they have closed a story, not opened one. Because it is predictable that such papers will seldom be cited, the "top" journals will often not consider publishing them.

If we propose a bold hypothesis, mostly we will be wrong. If we could be sure that it was not wrong, it would seldom be bold. But the scientific literature is not filled with refutations. It is hard to design and conduct killer experiments, and it is a high-risk business — the perfect experiment might destroy a persuasive narrative that the scientist has spent years in developing. So the literature is filled with confirmation and circumstantial support, sometimes with circuitous arguments that appear to deny what the evidence seems to say, sometimes with a creative use of statistics, sometimes by a creative recasting of a hypothesis. Data seldom speaks for itself, and it is sometimes tortured to say the strangest of things.

In 2010, Daniele Fanelli studied a large sample of scientific papers in areas of both pure and applied science—from physics and chemistry to psychology and social sciences. He analyzed papers that "by declaring to have tested a hypothesis, had placed themselves at the research frontier of all disciplines and explicitly adopted the hypothetico-deductive method of scientific inquiry, with its assumptions of objectivity and rigour."[9] The odds of reporting a positive result were higher for applied research than for pure research and higher for social sciences than physical sciences, but, in all fields, the great majority of papers reported evidence in favor of the tested hypothesis. At the top of the list, more than 90 percent of the papers in psychology and psychiatry reported positive findings.

Let us not get too carried away by this record of success. Put it like this, if scientists can correctly predict the outcomes of their experiments 80–90 percent of the time, why do they waste so much time and money actually doing them?

The conclusions seem inescapable. Either scientists misreport the outcomes of experiments as supportive when they are not. Or they misrepresent their hypotheses—the hypotheses that they purport to be testing were conceived not before the experiments but after them, to fit in with the outcomes. Or they misrepresent the state of understanding at the outset of their experiments: their experiments are not innovative leaps into the unknown, inspired by bold hypotheses as they are purported to be, but have outcomes that are predictable (and predicted) from evidence and reasoning hidden from the readers. Or they suppress inconvenient evidence: many experiments that fail to support their hypotheses are never reported in the scientific literature.

By any of these conclusions—and there is reason to think that all are valid—the hypothetico-deductive method, the supposed cornerstone of modern science, is a sham.

Scientists are remembered not for things they got wrong, but things they can claim to have gotten right. Thus, scientists accentuate the positive in the outcomes of their studies—they cleave to findings that are clear, and then construct a hypothesis according to which these particular findings should be expected, while minimizing or disregarding any findings for which they have no reasonable interpretation. Such narratives give the appearance of hypothesis-driven research, but this appearance is artificial. The research might have been driven by a hypothesis—but not by the hypothesis apparent from the narrative. The hypothesis as presented is an imposter.

The scientific paper of Medawar's time may indeed have been a fraud; but perhaps a benign fraud—the traditional narrative form made it easier for the reader to follow what the scientist had done. The contemporary scientific paper is a fraud in a more pernicious sense.

In the next chapter, we look at paradigm shifts: what happens when anomalies accumulate from experimental evidence that are beyond the ability of scientists to construct plausible explanations for them.

5 The Birth of Neuroendocrinology and the "Stuff of Legend": A Case Study of Paradigm Change

For all its rhetorical force, Kuhn's analysis of Science was constructed by inferences from selected episodes of what now seems ancient history. We might expect, from Popper, that scientific theories are constantly in turmoil. We might expect, from Merton, that scientists readily exchange knowledge and technologies through collaborations. We might also recognize that fields are not policed by rigid disciplinary demands—they are typically so broad that scientists are never constrained by a single paradigm but can migrate amongst several. So in what sense is a field *ever* in crisis, and is there ever a scientific revolution?

Latour and Woolgar noted that neuroendocrinology, the field described in *Laboratory Life*, seemed to have "all the attributes of a mythology: it had had its precursors, its mythical founders, and its revolutions."[1] The principal "mythical founder" of neuroendocrinology was understood to be Geoffrey Harris (1913–1971).[2] The "revolution," as Latour and Woolgar understood the myth, consisted in the overthrow of one understanding of how the pituitary is regulated (by hormonal feedback from its target tissues in the body) by a different understanding (that it is regulated by hormones secreted from the brain). "As in many mythological versions of the scientific past, the struggle is now formulated in terms of a fight between abstract entities such as models and idea."[3]

But, in the memories of those involved, the fight was a very personal one, between Geoffrey Harris and Solly Zuckerman.

The Harris–Zuckerman Debate

By the late 1950s, it was a "well-established fact" that, in mammals, ovulation is controlled by the anterior pituitary gland, which lies just below the brain.[4] In each ovarian cycle in a female mammal, one or more of the follicles that are present within the ovaries begin to develop. Each follicle contains a single ovum—an egg that, if fertilized by a sperm, will become a fetus. The follicles mature in response to the secretion from the pituitary of a "gonadotropic" hormone, and as they mature, they produce estrogen. The increasing levels of estrogen stimulate the secretion of more and more gonadotropin, and eventually this triggers ovulation—the release of the ova from the mature follicles—ending one ovarian cycle and beginning the next.

The common view in the 1950s was that estrogen stimulated gonadotropin secretion by its effects on the pituitary. But Harris argued that estrogen acted not there but on the brain, resulting in the secretion from the brain of some substance that was carried by blood to the pituitary.[5] In his theory, this substance was one of several *releasing factors*, each produced by a different population of neurons in the brain, each of which regulated a different pituitary hormone.

This theory had much to dislike about it. For those who favored parsimony in their theories, it engaged an unnecessary complication. It blurred the distinction between neurons, the stuff of the brain, and endocrine cells, which secreted hormones into the blood. It postulated the existence of substances—releasing factors—the identity of which was beyond speculation and seemed to be far beyond experimental reach.

But anatomists had already concluded that the anterior pituitary contained only endocrine cells, and they had seen that a system of small blood vessels at the base of the brain was connected to it. Harris showed that these vessels branched repeatedly to form a "vascular plexus" that permeated every part of the anterior pituitary, and that blood flowed in these vessels from the brain to the pituitary.

He then noted that when other endocrine glands (the testes, thyroid, parathyroid, and adrenal cortex) were transplanted to various parts of the body, they could function normally. But when the anterior pituitary was transplanted, it always stopped working—the peripheral tissues on which pituitary hormones act all withered away.

His 1955 monograph, *The Neural Control of the Pituitary Gland*, which became recognized as the moment of birth of a new scientific field, includes a sentence that Bruno Latour and Steve Woolgar might have appreciated:

> Since the statement that "anterior pituitary tissue transplanted to a distant site shows very little functional activity" is of such significance in the present discussion, some of the older work, which claimed that functionally active transplants had been obtained, will be examined in more detail.[6]

This sentence is followed by five pages that address all previously published reports of pituitary transplantation. There was no evidence that the anterior pituitary could function normally when transplanted to another part of the body. But if it was removed from its normal site and then replaced there, then normal function often did return. Harris hypothesized that in these cases, the hypothalamic blood vessels had regenerated, renewing the communication between brain and pituitary.

To test this, Harris and his colleague Dora Jacobsohn[7] removed the pituitary from female rats and replaced it with pituitary tissue from their newborn young. They grafted the tissue into the space below the brain, either immediately below the portal vessels or just to one side below the temporal lobe of the brain. By so doing, they could confirm histologically that the rats' own pituitaries had been completely removed, as well as establishing whether the grafted tissue was in fact revascularized, as they hoped it would be.

The tissue survived equally well at both sites. In both cases the transplants were revascularized by blood vessels from the brain—in the first case by the portal vessels, in the other by the blood vessels from a different part of the brain. Normal ovarian cycles returned in all twelve rats with transplants below the cut portal vessels, and six of these later became pregnant. But in the rats where the transplants were below the temporal lobe, the ovaries and reproductive tracts were atrophied and reproductive function ceased.

These were experiments in the Popperian mold: "Those among us who are unwilling to expose their ideas to the hazard of refutation do not take part in the scientific game."[8] If, in just one of Harris's rats, ovarian function had been restored by a transplant that received no blood from portal vessels, or if ovarian function was not at least partially restored by a transplant revascularized in this way, then Harris's bold theory would have been refuted. And it could still be refuted by anyone willing to repeat his experiments.

This was no distant threat, for Harris had engaged the opposition of Solly Zuckerman, an extraordinary man who then held the positions of professor of anatomy at Birmingham University and professor at large at the University of East Anglia. Zuckerman had founded his academic career on studies of the menstrual cycle in primates, but during the Second World War, he had reached a different level of recognition as scientific advisor to Combined Operations Headquarters. After the war, Zuckerman—soon to be Sir Solly Zuckerman, later to be Baron Zuckerman—combined his academic role with that of science advisor to successive governments. According to a former government minister, he was "one of the most influential figures in the nebulous and powerful network, sometimes called the establishment, which lies at the heart of much of Britain's national decision-making."[9] The Nobel Prize winner George Porter described Zuckerman's strengths as "the application of scientific method to every problem that confronted him, many of them not seen as scientific" and "the personal authority that made his advice count."[10] Lord Dainton, a former chairman of the UK Council for Scientific Policy, wrote: "Solly Zuckerman was unique. No scientist this century can match him in the timespan and weight of his influence on governments in peace or war."[11]

The Zuckerman–Harris controversy, in the words of Seymour Reichlin, a scientist who was trained by Harris, "is the stuff of legend."[12] Zuckerman recognized that if any animal could be shown to have ovulated in the absence of portal vessels, then Harris's theory would be refuted. The experimental animal that Zuckerman chose for his assault was the ferret. Ferrets are seasonal breeders—they are reproductively inactive during the winter months, but as the days lengthen in spring they come into heat and they ovulate.

Heat describes the phase of the ovarian cycle in which female mammals are sexually receptive, and in ferrets, this could be easily recognized by swelling of the vulva. This is a response to increasing levels of estrogen, and it can easily be induced by exposing ferrets, in the winter months, to artificial light.

In 1953, A. P. Thomson and Zuckerman reported that they had cut the connection between the brain and the pituitary in sixteen female ferrets, and that, after exposing the ferrets to artificial light, ten had come into heat.[13] Accordingly, the ovaries must have been producing estrogen, and for this to be the case, the pituitaries must have been functional. For eight of the ferrets they reported that there was evidence of some reconnection of the blood vessels with the pituitary, but in two ferrets they declared that all connections had been eliminated. Zuckerman's case for refuting Harris's bold hypothesis rested on these two ferrets.

Harris visited Zuckerman's laboratory to examine the evidence; he accepted that the connection—the *neural stalk*—had been cut in the two ferrets concerned, but he suspected that the portal vessels had regenerated. With a PhD student, Bernard Donovan, Harris carried out a replication study.[14] In one group of ferrets, they cut the stalk in the manner of Thomson and Zuckerman. In another group, after cutting the stalk, they inserted a paper plate between the stalk and the pituitary to prevent any revascularization. In most of the first group there was revascularization of the pituitary (which they showed by using what they claimed were better techniques than those used by Zuckerman), and several of these ferrets ovulated in response to light. In the second group, in all cases where the plate had been inserted effectively, there was no revascularization and no ovulation.

Harris did not dispute that Thomson and Zuckerman had seen what they said they had seen. He accepted that ten of the ferrets had ovulated, and that the stalk had been cut in these ferrets. He also accepted that if these cuts *had* eliminated the blood supply to the pituitary, then Zuckerman was justified in concluding that his theory was wrong. But he concluded that Thomson and Zuckerman's experiments were technically flawed.

The argument came to a head in 1954 at a conference in London at which Harris and Zuckerman spoke. In Reichlin's memory of that debate, Zuckerman was "by far a more cunning and skillful debater and spoke with deliberate and practiced authority summarizing his objections to Harris's theories," but "Harris won the debate, then, and in posterity."[15] Harris won not only because his evidence was more convincing but because he could explain something that Zuckerman could not. Zuckerman's two ferrets had come into heat in response to light—hence the pituitaries had responded to signals from the retina despite what Zuckerman had claimed to be a complete separation of the pituitary from the brain. For this, Zuckerman had no credible explanation. In his own words, "Our experiments do not give any indication of the way in which the response was mediated."[16] This absence of a mechanistic explanation critically undermined his argument.

Throughout this debate, we might recognize Merton's norms: communism, in Zuckerman's willingness to show Harris his evidence; universalism, in the collective judgement of the merits of the argument by a community of peers; disinterestedness, in the independence of that judgement on the relative status of the participants; and organized skepticism, in the processes of challenge and replication. We can recognize too the application of Popper's principle of falsification.

We can also find confirmation of Kuhn's insight that scientists typically cling tenaciously to their beliefs in the face of contrary evidence. Zuckerman still maintained his position in 1978, even though Andrew Schally and Roger Guillemin had, the previous year, been awarded the Nobel Prize for the chemical identification of the very factor that Harris had postulated stimulated the secretion of gonadotropins, the factor whose existence Zuckerman had disputed.[17]

Zuckerman's persistence seems bizarre: "To the community of neuroendocrinologists, Sir Solly's implacable pronouncements have been inexplicably equivalent to the rejection of Galileo's view of the relationship between the Sun and planet Earth."[18]

To understand it, we might recognize that Harris's theory posed a threat to Zuckerman's status as the foremost expert on the reproductive cycle. But Harris's theory posed a broader threat. Both Zuckerman and Harris were, by training, anatomists. Anatomical observations gain their significance through their interpretation. They are interpreted in the light of experiments that involve interventions and experiments that elucidate mechanisms. For Zuckerman's studies of the reproductive tissues, the relevant experiments were those of endocrinologists, who applied specialized techniques to study hormonal communication between tissues. Harris, however, placed the brain as the controller of reproduction, and the consequence of so doing was to shift the responsibility for extending our understanding to neuroscientists, who employed radically different techniques, such as electrophysiology. In this, Harris posed an existential threat to the community of reproductive scientists of which Zuckerman was the preeminent champion, a community not technically equipped to adapt to this change in focus.

That this threat was real soon became apparent. A new field, neuroendocrinology, was born, soon with its own journals and later, as the schism from endocrinology deepened, with its own professional societies. One of the new neuroendocrinology laboratories was that of Roger Guillemin, the lab in which Bruno Latour had spent two years as an anthropologist of science; *that* lab was devoted to the identification of the releasing factors that Harris had predicted.

Guillemin's eventual success in this led him to a share, along with Schally, of the 1977 Nobel Prize in Physiology or Medicine. Harris's early death had denied him any prospect of a share in that prize. In 1971 Schally had written to Harris expressing his regret that Harris had not won the Nobel Prize of that year and predicting that his time would come soon; the letter arrived a day after Harris's death.[19]

Harris's students and disciples—Seymour Reichlin and Bernard Donovan amongst them—scattered and seeded the new field. Another of them, Barry Cross, who pioneered the integration of electrophysiological techniques into neuroendocrinology, later reflected on the Zuckerman–Harris controversy:

> Of course, it is much easier to be critical of other people's theories than their own. The point is that all theories are suspect until tested and retested, until destruction if necessary. Harris needed his Zuckerman. What endures from such controversy approaches more nearly the truth.[20]

The Paradigm Shift of Neuroendocrinology

We can recognize in the Harris–Zuckerman controversy what Kuhn might have called a scientific revolution—a paradigm shift. For Kuhn, a new paradigm had two defining features: it was novel enough to attract followers away from other modes of scientific activity and open-ended enough to leave ample problems for those new followers to resolve.[21] He elaborated on this in four ways:

1. "The paradigm sets the problem to be solved; and often the paradigm theory is implicated directly in the design of apparatus able to solve the problem."[22] Harris's theory set the key

problem as to identify the releasing factors, and it is this that engaged the energies of Guillemin and Schally. At the same time, the theory raised the questions of those factors were made and how they were regulated—open-ended problems that demanded new approaches.

2. "The formation of specialized journals, the foundation of specialists' societies, and the claim for a special place in the curriculum have usually been associated with a group's first reception of a single paradigm."[23] Neuroendocrinologists founded new journals—including *Neuroendocrinology*, *Frontiers in Neuroendocrinology*, and the *Journal of Neuroendocrinology*. They formed professional societies in many counties, societies that now come under the umbrella of the International Neuroendocrine Federation. Neuroendocrinology entered the undergraduate curricula in medical and biomedical sciences.

3. "During the period when the paradigm is successful, the profession will have solved problems that its members could scarcely have imagined and would never have undertaken without commitment to the paradigm. And at least part of that achievement always proves to be permanent."[24] Synthetic versions of the releasing factor for gonadotropins are now an indispensable adjunct to *in vitro* fertilization, a process by which about 1 million babies were born in the United States alone between 1985 and 2012.[25]

4. "Crises are a necessary precondition for the emergence of novel theories."[26] A crisis in science, for Kuhn, is occasioned by the accumulation of "severe and prolonged anomalies." The theory championed by Zuckerman was that the ovarian cycle arises through hormonal interactions between the pituitary and the ovaries. However, this failed to explain how, in some species, ovulation is triggered by sexual activity, or

how, in some other species—including in ferrets—ovulation occurs at a particular time of year. Zuckerman's suggested answers were that, in the first case, the pituitaries of those species might receive some innervation from the brain. But in the second case his own experiments had exemplified the anomaly—his two ferrets ovulated in response to light despite what he had claimed to be a complete separation of the pituitary from either nerves or blood vessels. For this, he had no credible explanation.

While Harris had claimed to have better evidence—a better way of cutting the neural stalk and of visualizing blood vessels—few of his audience had the technical expertise to evaluate the merits of this argument. The debate was won by Harris because he had done enough to cast doubt on Zuckerman's and because his account could explain what Zuckerman could not—that the pituitaries of his two ferrets could still respond to signals from the retina. As Kuhn observed: "The decision to reject one paradigm is always simultaneously the decision to accept another, and the judgment leading to that decision involves the comparison of both paradigms with nature and with each other."[27]

Rivals

Any case study such as this is colored by hindsight, omitting many diverse factors and influences. While we highlighted the apparent observance of Merton's norms, we might equally have documented many breaches. Schally and Guillemin, so far from being brothers in a common enterprise, were jealous rivals, in pursuit of the most prestigious prize in science. The story of their rivalry has been well documented: Schally described it as involving "many years of vicious attacks and bitter retaliation." They cited as little as possible of each other's work. Schally refused to share his materials with Guillemin: "It is like giving someone a gun so he can shoot you. What did he make so much fuss for? He was an opponent and an enemy at that time."[28]

The actual identification of the releasing factors was less an intellectual challenge than the application of "brute force." If enough hypothalamic material was available, processes were readily available by which they could be isolated. What both Schally and Guillemin did was collect hypothalamic material on a massive scale: Guillemin's lab, in a few years, collected more than fifty *tons* of hypothalamic fragments from 5 million sheep brains. Supported by several pharmaceutical companies, both Guillemin and Schally gathered massive resources, of both money and people, and deployed them in a single-minded pursuit of a Nobel Prize.[29] The work of isolating the releasing factor had little theoretical interest: Cy Bowers, one of the biochemists harnessed by Schally, commented, "Most intelligent people won't do isolation work—I think my IQ went down about 20 points while I was doing it."[30] It seemed that the Nobel Prize had become a lottery, in which Guillemin and Schally had bought all available tickets, with nobody left to beat except each other. Neither seemed happy that the award was shared between them.

The rivalry between Schally and Guillemin might seem to have involved a profligate duplication of effort and resources. Now, such duplication is discouraged by grant funding agencies, and although it has been a guiding principle that independent replication of experimental findings is of key importance in science, journals are reluctant to publish replication studies and funding bodies are even more reluctant to fund them. However, that Schally and Guillemin independently came to the same conclusion about the nature of the releasing factor for gonadotropins provided compelling support for their claims, and subsequent progress was swift.

The rivalry had another important consequence. In this race, any error by either lab in either the technical processes or in the interpretation of data had the potential to delay progress if not derail it altogether. Furthermore, any claim of any substantial importance would, if subsequently contradicted, diminish the credibility of the lab that generated it. Given the intense competition, the likelihood of error being exposed was considerable. Such rivalry is a check on the integrity of science: scientists who aspire to claims of importance and who hope to sustain a lasting career in science protect themselves against the risk of error by honesty and by self-criticism.

Merton had characterized the norms of science as moral precepts:

> The ethos of science is that affectively toned complex of values and norms which is held to be binding on the man of science. The norms are expressed in the form of prescriptions, proscriptions, preferences, and permissions. They are legitimatized in terms of institutional values. These imperatives, transmitted by precept and example and reinforced by sanctions are in varying degrees internalized by the scientist, thus fashioning his scientific conscience.[31]

We might now understand these norms as amoral consequences of scientists being, in Latour's words, "in the business of being convinced and convincing others."[32] Their long-term credibility and the career advancement that this brings depend on the reliability of the evidence that they present, on the scope and importance of their claims, and on the strength of those claims as measured by their ability to withstand critical assault. They are, for the most part, honest in their science, simply because honesty is the best safeguard against error. But they are not paragons.

6 The Language of Crisis and Controversy, and the Levers of Paradigm Change

Neither the story of Harris and Zuckerman nor that of Schally and Guillemin are representative examples of science. Few scientists have the intellectual gifts of Harris or the establishment status of Zuckerman, few are credible candidates for a Nobel Prize, and vanishingly few seek it with the avidity of Schally or Guillemin.

From Latour's account of life in the laboratory, we can recognize that there is continual questioning, skepticism about the validity of the apparent results of experiments and about the published results of others. Generally, these doubts are expressed in the most genteel of ways. The lay reader of a typical scientific paper might scarcely register the differences between "it has been reported that" and "it has been established that," or between "it has been claimed that" and "it has been concluded that"—but to those directly concerned, these subtle differences are telling and can be inflammatory. They tell of the disputes and dissension that are part of the fabric of science. However, sometimes these gentilities give way to open warfare, and by this we might recognize times of crisis.

The Wounded Lion

When Harris published his monograph, Zuckerman reviewed it for *Nature*, and his scorn for it was undisguised. His opening words set the tone: "The place taken by the hypothalamus in current biological discussion is in striking contrast to its size. This minute region of the brain also remains a happy hunting ground for experimentalists concerned to find in it neural 'centers' for almost every visceral function of the body."[1]

He went on to state that "Prof. Harris's views deepen rather than lighten the mystery that surrounds the physiology of the hypothalamus." He listed some of "the many things that the critical reader needs to be told, but is not," and bewailed his "speculations about the number of chemotransmitters that might be involved (when not even a single one has been identified)."[2]

In the following years, every one of the chemotransmitters that Harris had postulated was identified. They now form the basis of clinical treatments for a wide range of endocrine disorders. Each was found to be made by a particular small population of neurons in the hypothalamus, a region that turned out to have far more functions than were ever imagined at the time of Zuckerman's review.

Zuckerman's were the words of a wounded lion, raging "against the dying of the light." But what about the young lions, those who challenge accepted wisdom?

The Young Pretender

In 1981, an article from a young biochemist, Tony Trewavas, provoked a controversy in plant biology.[3] At that time, his field was dominated by the study of plant hormones, factors released by cells of a growing plant that guided its growth and development. Scientists sought to identify these factors, to study their effects on growth and development, and to measure their abundance in particular plants in different conditions. The paradigm was dominated by a view formed in analogy with what was believed to be true of animal hormones, that regulation by hormones of a physiological response is exerted by control of the rate of production of those hormones.

Trewavas argued that the analogy was weak: animal cells had separated from plant cells before the emergence of complex multicellular organisms, so intercellular communication must have evolved independently in plants and animals. Without a common evolutionary origin, there was no good reason to think that plant hormones and animal hormones worked in similar ways. Indeed, there was good reason to doubt this: whereas animals had a medium of communication—blood—that could deliver hormones from their site of production to distant sites of action, plants lacked any obviously analogous system. Secondly, communication between cells involves two things—a messenger and a receptor for that messenger, something that specifies the sensitivity of cells to the message. Changes in growth and development that had been attributed to changing levels of a plant hormone might be equally well explained by changes in the sensitivity of plant cells to *unchanging* levels of the hormone. Thirdly, Trewavas catalogued examples where measured hormone levels did not correlate with changes in growth or development, and he disparaged the evidence for claims to the contrary.

Trewavas began his article by portraying a crisis: "Despite 50 years intensive research input on plant growth substances there can be few who find particular satisfaction with the current state of the field. Those who work in the area will be only too familiar with the often-confusing contradictions, the apparently endless and puzzling interactions and the plain uncertainties of supposedly established facts."[4]

He continued by detailing contradictions ("The only occasion on which this has been measured has, in fact, shown the opposite") and errors in the interpretation of data ("The circular nature of this argument should not escape the reader"). He also detailed methodological shortcomings: "The apparent linearity is deceptive. It seems to have arisen from use of a very low range of auxin concentrations, a failure to apply proper statistical procedures, and the use of an insufficient number of plants to eliminate the variability inherent in the method."[5] In these arguments, Trewavas observed the convention that, in critical analysis, it is the message that is attacked, not the messenger. But he went further, in an assault on the scholarship of his more senior contemporaries in the field:

> "It is difficult to see why these views have prevailed for so long when they had such obvious errors."
>
> "The error [. . .] is a historical one but there is no reason why past mistakes should continue to be repeated."
>
> "Those who work on plants have tended to apply many of these ideas rather uncritically to their chosen organism."
>
> "Although these [properties of plants] may be familiar to the reader, if awareness is measured by research interest then the current state of awareness is abysmally low."[6]

These attacks, though not addressed explicitly to individuals, were nevertheless hurtful. Trewavas was attacking the credibility of eminent scientists, their scholarship, the stuff of which their self-esteem and their esteem in the eyes of their peers was constituted. He paid a price in things that mattered to a still junior scientist. His promotion was blocked by a hostile referee, and an invitation to publish a book chapter following a conference at which he had been invited to speak was withdrawn when other senior authors objected.[7]

But Trewavas had not simply attacked the consensus view—he had also presented an alternative, and because of this his arguments could not be ignored. Before the internet, scientists typically collected reprints of papers by direct requests to the authors, and he was soon getting more than sixty such requests a day, swiftly exhausting his supply. Journal editors noticed the interest—they like controversial articles for the citations they bring to the journal—and soon, Trewavas was getting invitations to extend his arguments.

One journal presented the debate as a face-off between Trewavas and one of the senior figures in the field, Bob Cleland.[8] Cleland presented the conventional view and dismissed Trewavas, stating, "Are changes in sensitivity to plant hormones also important in the control of plant development? Of course! This is hardly a new idea."

Trewavas was given the last word:

> Bob Cleland is aware of the importance of changes in sensitivity in controlling development. Good. But is his perception widely shared? If it is so well appreciated why is there so little published research on sensitivity? Why is it that in scouring the literature of the last thirty years I have found no more than a handful of examples which show any appraisal and systematic measurement of sensitivity changes during development? Why is there no reference to sensitivity in recent and, indeed, less recent textbooks if its importance was so obvious? [. . .] We have many hundreds of plant physiologists who frankly just make measurements of concentrations and less than a handful who investigate sensitivity.

The answers were obvious. The alternative view that Trewavas had articulated was unpalatable because pursuing it implied a methodological retooling of the field. It required studying the behavior of individual plant cells and their intracellular signaling mechanisms, things for which the expertise of most of his contemporaries in the field was unsuited.

To this end, Trewavas threw his energies, as did many newcomers to the field, stimulated by the controversy. The key intracellular signaling pathways involved calcium, and ways of studying this were well established in animal cells. These techniques were not easily transferred to plant cells, which are protected by a tough wall, impenetrable by the electrodes used by animal cell biologists, but by 1991 Trewavas had bypassed this impediment by generating transgenic plants that expressed aequorin, a protein derived from a species of jellyfish that emits flashes of blue fluorescence. The fluorescence is triggered by an interaction between calcium and aequorin, and so, when aequorin is expressed in plant cells, the levels of calcium can be inferred from the level of fluorescence emitted.[9] Soon Trewavas was elected a fellow of the Royal Society, the highest accolade available to scientists in the United Kingdom.

By 1991, in the words of one reviewer: "The efforts of A. J. Trewavas to stimulate a reappraisal of the role of sensitivity as a controller of response [. . .] at first met with extreme resistance, but it cannot be denied that he achieved his aim."[10] His strategy had been one of high risk. He had been fortunate in persuading an editor to publish that first paper, and he might easily have been frozen out by his contemporaries.

The tone of that first paper is not the normal tone of disputes in science. These are not the nuanced phrases that, for Latour, characterized the promotion of one scientific claim at the expense of another. Each of these nuanced phrases embodies a different judgement of credibility, differences that are recognizable to those intimately concerned, but which are often invisible to the casual reader. A scientist working within a paradigm must gather allies and build networks of colleagues who are willing to at least acknowledge his or her contributions, and the tempering of criticism is a necessary adjunct. Caustic attacks, when they do come, often come from the young and reckless, like Trewavas, or from the aged and distinguished, with nothing to lose, frustrated at what they perceive as a swelling tide of nonsense.

The Social Structure of Scientific Revolutions

For Kuhn, a paradigm was far more than just a theory. Most theories are in constant flux; they are repeatedly revised and are often reformulated in ways that make them virtually unrecognizable when compared to earlier versions. To understand the nature of scientific revolutions and why they are resisted so strenuously, we must understand the social structure of science.

Every scientist invests heavily in the acquisition of particular technical expertise over long periods of training and acquires credibility by using that expertise to generate claims that are relevant and influential. Paradigm change does not necessarily occur when a theory is overthrown, but when the expertise of a community loses its relevance, in being unsuited to answering new questions that arise from a new theory, a new discovery, or a new technology. This practical consideration, with all its implications for prospects of career advancement, wins adherents for the new paradigm at the expense of the old. The new recruits are not necessarily attracted by the intellectual superiority of a new paradigm, but make a pragmatic decision to invest in skills that will be in increasing demand.

We might see, from the Harris–Zuckerman debate and the Trewavas controversy, the conditions in which a crisis does arise. In both cases, the crisis arose because of an existential threat to a community of scientists, a community that, because of its insularity, was technically ill-equipped to respond to it. As a result, that community was threatened with a loss of credibility, a dearth of new recruits, and a loss of competitive funding. *Of course* the challenges were vigorously resisted; *of course* scientists were reluctant to change their minds—they had no real alternative but to resist. The revolutions that ensued were not so much scientific revolutions, although they did entail theoretical adjustments; they were social revolutions, a wholesale reconstruction of the social structures of those communities—the networks of collaboration, technology exchange, and citation.

But how does this idea of how scientific understanding progresses fit with philosophical views of the rational nature of science?

7   Logical Positivism: The Trouble with Verification

Le doute n'est pas une état bien agréable, mais l'assurance est un état ridicule. (Doubt is not a pleasant condition, but certainty is an absurd one.)

—Voltaire[1]

When Popper was writing *The Logic of Scientific Discovery*, the dominant movement in philosophy was *logical positivism*—the view that the only meaningful statements (other than the purely deductive statements of mathematics and logical tautologies) were those that could be verified empirically.[2] This saw science as a progressive activity, building from empirically demonstrable "facts," drawing inferences from them, and verifying those inferences by further empirical studies. Logical positivism, in its emphasis on the empirical elements of knowledge, was an attack on "woolly thinking" and on exaggerated respect for authorities. It gained wide traction through the work of the philosopher A. J. Ayer (1910–1989), particularly through his book, *Language, Truth, and Logic*.[3] Written when Ayer was just twenty-four years old, this book reached a wide audience through its brevity and clarity. Thelma Zeno Lavine, professor of philosophy at The George Washington University, wrote in 1983 that Ayer was "an enfant terrible who cleverly placed a lighted stick of dynamite under all traditional philosophies. The old philosophic landscape has never been fully rebuilt since then." His book, she continued, "is generally conceded to be one of the most influential books of 20th-century philosophy."[4]

But in 1967, John Passmore concluded his entry on logical positivism in *The Encyclopedia of Philosophy* with the words: "Logical positivism, then, is dead, or as dead as a philosophical movement ever becomes."[5]

The Vienna Circle

Ayer, while still a student, had traveled to Austria to study the works of the Vienna Circle, a group of philosophers, mathematicians, and scientists who debated the relationship between evidence and conclusion and logic and experience. Their debates engaged four questions:

1. What relationship should philosophy have with the empirical sciences?
2. What relationship exists between the empirical science and logic and mathematics?
3. What relationship exists between evidence and conclusions?
4. How can empirical knowledge be justified?

For Moritz Schlick (1882–1936), the chair of the Circle, to answer these questions philosophy must first to be stripped of its metaphysical pretensions:

> The only fruitful method of all theoretical philosophy consists in critical inquiry into the ultimate principles of the special sciences. Every change in these ultimate axioms, every emergence of a new fundamental principle, must therefore set philosophical activity in motion. [. . .] It is primarily, or even exclusively, the principles of

the exact sciences that are of major philosophical importance, for the simple reason that in these disciplines alone do we find foundations so firm and sharply defined, that a change in them produces a notable upheaval, which can then also acquire an influence on our world-view.[6]

Schlick thus argued that philosophers of science should not agonize about the existence of the real world. Their role was not to question scientific facts, but rather to understand the processes by which scientists inferred them. The Circle proposed to achieve this by analyzing the structure of scientific language—to use philosophy to unveil the hidden logical structure to this language and its correspondence with reality.

The manifesto of the Circle set out its goal:

Neatness and clarity are striven for, and dark distances and unfathomable depths rejected. [. . .] Clarification of the traditional philosophical problems leads us partly to unmask them as pseudoproblems, and partly to transform them into empirical problems and thereby subject them to the judgment of experimental science. The task of philosophical work lies in this clarification of problems and assertions, not in the propounding of special "philosophical" pronouncements.[7]

By the Circle's understanding, scientists methodically experiment with nature, record their observations systematically, and by this process generate robust, immutable facts. A principle of *empirical verification* underpinned the validity of those facts. As Ayer expressed it, "A proposition is said to be verifiable, in the strong sense of the term, if, and only if, its truth could be conclusively established by experience."[8]

Sociologists shared this narrative. According to Merton, the aim of science was to extend "certified knowledge":

The technical methods employed toward this end provide the relevant definition of knowledge: empirically confirmed and logically consistent statements of regularities [. . .] the technical norm of empirical evidence, adequate and reliable, is a prerequisite for sustained true prediction; the technical norm of logical consistency, a prerequisite for systematic and valid prediction.[9]

Historians too once shared this narrative, now contemptuously called *Whig history*. They presented the history of science as a story of progressive refinement, a march to monistic truth accomplished by the ever-increasing resource of "facts." The present state of science was seen as always better than that which had gone before. Deviant facts and theories that did not fit with current knowledge were the result of errors of reasoning; older theories were reconstructed and reinterpreted to cohere with present day knowledge and used as exemplars of the diligent application of reason and of the scientific method.

Thus, when logical positivism collapsed, it brought down with it a narrative that had been shared not just by philosophers and scientists, but also by sociologists and historians. The collapse came not from an attack on the narrative, but from the failure of an attempt to make it rigorous.

The logical positivist had set out to explore what an objective scientific language would look like, and how it might lead to "justified belief" in a fact.

Protocol Sentences

Logical positivists recognized that scientists propose hypotheses that could be tested against the facts of empirical experience. They proposed that these facts could be expressed in *protocol sentences*, base units of knowledge that report direct sensory experience. Only hypotheses that implied conditions whereby they could be rendered either true or false by protocol sentences were permitted—they had to be *verifiable* to be meaningful.

But this encountered logical problems: "Only statements that are verifiable are meaningful" is not a scientific statement, so should *it* be rejected as meaningless? Logical positivists argued that this was not a statement but a *prescription*. This was an uncomfortable position to hold, given that they had set out to eliminate metaphysics from the philosophy of science.

Their attempts to reduce scientific statements to their objective essence were not without difficulty. It required engaging with problems of subjectivity (I see this as red, but do you?) and methodological protocol (when and in what conditions was an observation made). The reliability of a sensory observation depended on many things: on the observer being honest and unbiased, and on the future being like the past. If our senses and reasoning are fallible, if how we describe those experiences is subjective, and if observations are subject to the vagaries of chance and circumstance, how can the truth of an observation ever be established? Different members of the Circle struggled to see how protocol sentences might be formulated to circumvent these issues. Otto Neurath (1882–1945), one of the Vienna Circle's champions of the scientific attitude, proposed a "paradigmatic example" of a protocol sentence, an example that displayed the hopelessness of the task:

> Otto's protocol at 3:17 o'clock: [Otto's speech-thinking at 3:16 was: (at 3:15 o'clock there was a table in the room perceived by Otto)][10]

Schlick saw, to his horror, that incorporating the subjectivity of the observer into protocol sentences left open the possibility that they might not be true, rendering unsound any deductions made from them.[11]

Ludwig Wittgenstein (1889–1951) in his later work cut through this nonsense: he denied that language assembles units that have any objective, immutable correspondence with reality.

The meaning of words depends on the context in which they are being used.[12] Words are part of the means by which a speaker attempts to impart an idea from his or her mind into the mind of the hearer, but they can never fully express that idea, they always rely on assumed shared knowledge, they always depend on context, and the idea will always be changed in the process of transmission.

Justified Belief

Logical positivists sought to understand the logical structure that they presumed to underlie the facts of science.

In any purely deductive system, the truth of a statement is an outcome of how we define concepts and the rules of correspondence that we establish. Thus, in deductive systems, anyone who understands the meaning of terms and who follows the rules must come to the same conclusions. For example, if we assert that A is an example of the class, X, and B is also an example of X, then A and B are both examples of X—it is a tautology.

In such deductive systems, the truth of statements always depends on how we define the terms and rules. But the logical positivists did not accept that the truth of scientific statements depended on definitions in quite the same way. Observation statements—the basic facts from which other facts were constructed—were not arbitrary: they were derived directly from sense data, which could be assumed to be the same for all able observers.

A scientific "fact" about observation statements, such as the claim that leptin inhibits appetite, acquired *its* meaningfulness by the protocol used to verify it (including how to measure leptin and appetite), and once verified, its truth was also resistant to changes in conceptual frameworks. The terms "leptin," "inhibit," and "appetite," if given precise operational definitions, have an enduring meaning that is anchored to reality by empirical observables. Thus, while "facts" might be expressed in different ways in different conceptual frameworks, their truth survived redefinition because of their ultimate dependence on observational statements. Accordingly, older scientific theories could be reconstructed to cohere with modern theories because their basic observation units (if they had been properly verified) remained valid. Theories might change, but the "facts" of science survive.

This allowed the logical positivists to justify belief in scientific knowledge. Scientific knowledge was inductively inferred from empirical observations and was the product of *justified true belief*—beliefs that were empirically verifiable and logically consistent.

But logical positivism relied on the notion that the truth of a statement could be established by verification. Verification—making repeated observations of a phenomenon—is a case of reasoning by induction. This is the process that Popper had declared to be logically invalid as a method of establishing truth. And although many experiments confirmed that leptin inhibits appetite, in the most relevant circumstance it does not: obese people typically do not respond to leptin—they are "leptin-resistant."

Carl Hempel (1905–1997) added fuel to the bonfire with his "paradox of the raven."[13] Consider the hypothesis that "all ravens are black." We might imagine that we can gain evidence of its truth by seeking out ravens and noting their color. But this hypothesis is logically equivalent to the hypothesis that "everything that is not black is not a raven." *This* we should be able to verify by seeking out things that are not black and noting that they are not ravens. Because the hypotheses are logically identical, it follows that by sitting in a meadow and recording examples of green and yellow things that are not ravens, we should gain confidence in the proposition that all ravens are black.

Logical positivists came to realize these problems. Ayer conceded that statements of science that are derived from a finite set of observations can never be verified conclusively. He proposed a "weak" form of verificationism instead—that a statement "is verifiable in the weak sense, if it is possible for experience to render it probable."[14]

In this form, even the logical positivists recognized that logical positivism was empty; one by one they conceded and left the stage. With Ayer's concession to Popper, logical positivism had forfeited all claims to any coherent logical foundation.

In a televised interview in 1972, Ayer was asked "what in retrospect do you think the main defects of this movement [logical positivism] were?" Ayer's response was ". . . the main defect was that nearly all of it was false."[15]

8   Ambiguity of Scientific Terms

I have chosen the title *Aspects* because it is unscientific and vague, because it leaves us the maximum of freedom, because it means both the different ways we can look at a novel and the different ways a novelist can look at his work.

—E. M. Forster[1]

Despite its problems, logical positivism might be the best description of what scientists themselves think that they do. Scientists, most of the time, think that there is a real world, that there are facts that can be known, that Science accumulates factual knowledge, and that it progresses because it follows "rational" principles of reasoning. Logical positivists admired how scientists strove for clarity and precision, and the efforts that they had made to eliminate metaphysical concepts from psychology and "vitalistic" concepts from biology.

As seen through the eyes of the logical positivists, scientists develop sophisticated technical languages that assign discovered or invented "things" names that distinguish them precisely from other "things." They then classify these into categories based on their similarities with other observed "things." They develop "operational definitions" by which this classification can be made objective and reproducible. This clarity and precision seemed to be important not just for individual scientists in their own experiments, but also for communicating with other scientists.

However, some concepts such as "consciousness" and "intelligence" seem to be indispensable despite being beyond any robust operational definition. As Tony Trewavas noted, there is no agreed definition of intelligence.[2] Dictionaries conventionally define it from the anthropocentric view that only humans can be intelligent. But by definitions that focus on advanced reasoning and tool use, crows and parrots are intelligent: they can solve some problems faster than some humans can. A bee colony seems capable of cognition in much the same way that a person is: the colony collects information about its surroundings, remembers it, combines it with information about its internal state to make decisions that advance or protect its well-being, and it does all this with no single controlling brain. Intelligence is often defined by the "capacity for problem solving." By this definition, intelligence for any species must be described in the context of that species—in the context of what problems it encounters and are important to solve. Trewavas pursues this to consider whether even plants are intelligent. It might seem obvious that we should regard primates as more intelligent than plants, "But once we can identify how well an individual plant performs '. . . in its own particular environment and enables one species to dominate and exert power over other species . . .' [. . .] this conclusion might need to be reassessed."[3]

Stress

Many other concepts like "hunger," "depression," "motivation," "fear," and "reward" arise not only in the psychological literature but also in diverse contexts in the biological literature where they are used with a plethora of ad hoc operational definitions. As noted elsewhere, "stress," which Hans Selye first recognized as a physiological concept, defied his best attempts to define it coherently. One critic summarized his attempts: "Stress, in addition to being itself, was the cause of itself and the result of itself."[4]

The word stress in a physiological context was borrowed by analogy from physics—the notion of stress as applied, for instance, to a beam that may cause it to buckle. But we also use stress in common language to express very subjective feelings. What one person experiences as stressful, another may be indifferent to or regard as stimulating, and what is stressful in some circumstances may be innocuous in other circumstances. We cannot define stress by what causes it—the cause, a "stressor," can only be recognized by its effect, the "stress."

Thus, physiologists tried to define stress in terms not of the stimuli that evoke it but by the physiological responses involved and by the pathways that mediate those responses. As they conceived stress, living organisms survive by maintaining a steady state, a *homeostasis* that is constantly challenged by intrinsic or extrinsic forces (stressors). Both physical and emotional stressors trigger central and peripheral responses. Centrally, neural pathways are activated to stimulate arousal and focused attention and to suppress functions such as feeding and reproduction—to meet immediate needs at the expense of future needs.

This *general adaptational response* exhibits consistent features. The pituitary gland secretes *adrenocorticotrophic hormone* (ACTH) that stimulates the adrenal glands to produce glucocorticoid hormones. Glucocorticoids have effects on many tissues. They raise blood sugar levels, making energy available for use by our muscles. They reduce inflammation that might distract us and impair our ability to use those muscles. They act on the brain to influence memory—we remember events that are stressful. The sympathetic nervous system is also activated, increasing heart rate, respiration, and blood pressure.

An important component of this system is *corticotropin-releasing hormone* (CRH), one of the "releasing factors" postulated by Geoffrey Harris (see chapter 5). CRH controls the secretion of ACTH but also acts within the brain; there, it enhances arousal and fear responses and inhibits feeding.

So what is stress? We cannot define it by its causes—whether a stimulus is stressful depends on the individual and on the particular context. There are also many different types of stressors—including pain, hunger, emotional stressors, and constitutive stressors such as infection. Can we define stress by its consequences? The most robust of these is the increase in glucocorticoid secretion. But the normal secretion of glucocorticoids follows a circadian rhythm: in humans, the levels are highest in the morning when we wake, to increase energy availability for the new day. So we must talk of abnormally raised levels, with some criteria to decide what is normal for a particular individual. However, even so, if we define stress by the glucocorticoid response, this leads to the absurd conclusion that we can eliminate stress by removing the adrenal glands.

Nevertheless, can we use this as an operational definition of stress in experimental studies? The problems do not disappear: in pregnant animals and in lactating animals, the secretion of glucocorticoids is attenuated—in these states, energy supplies cannot be redirected to the muscles without harmful consequences for the fetuses or offspring. Can we really interpret the attenuated secretion of glucocorticoids as meaning that pregnant or lactating animals experience less stress?

If we cannot define stress by its causes or consequences, can we define it by its mediator? The same problem arises—CRH might mediate the stress response, but it seems absurd to conclude that we can eliminate stress by eliminating CRH. That would seem to be to kill the messenger because we don't like the message.

Thus, stress remains "a ghost in the machine." It's a troublesome ghost, because whenever even physiologists talk to physiologists about stress, they cannot escape the connotations that the word evokes from its everyday usage. Perhaps they might have coined another term altogether? But scientists need also to communicate to the broader scientific community and to the public—and they will still need to translate any newly coined jargon into terms that can be understood more widely.

Ambiguity

Perhaps concepts can be useful precisely in being ambiguous. Their flexibility allows operational definitions to be constructed pragmatically and to evolve with new knowledge. In many areas of science, only concepts that are effectively dead have fixed and immutable definitions. Ambiguous concepts can be useful in communication because they tap into a body of implicit knowledge in the receiver's mind, a body of knowledge that cannot be fixed but is in continual flux, and this fuzzy communication is a source of creativity in science. Not everything can be left ambiguous, but ambiguous terms signal the gaps in understanding that lie between certainties, and invite us to explore them.

Otto Neurath came to see this feature of scientific language, leading him to reject the notion of protocol sentences:

> We are like sailors who have to rebuild their ship on the open sea, without ever being able to dismantle it in dry-dock and reconstruct

it from its best components. [. . .] Imprecise "verbal clusters" are somehow always part of the ship. If imprecision is diminished at one place, it may well reappear at another place to a stronger degree.[5]

Scientific terms could not always be made precise; vague terms could not always be removed or redefined without unintended consequences for the whole theory.

There are clear limits to this. In much of chemistry and physics, eliminating vagueness is crucial. But even here, as the physicist Richard Feynman put it, "The real problem in speech is not precise language. The problem is clear language. The desire is to have the idea clearly communicated to the other person. It is only necessary to be precise when there is some doubt as to the meaning of a phrase, and then the precision should be put in the place where the doubt exists. It is really quite impossible to say anything with absolute precision, unless that thing is so abstracted from the real world as to not represent any real thing."[6]

Myocardial Infarction

For an example, consider the assertion that consumption of saturated fats is a cause of *myocardial infarction* (commonly known as a heart attack), a condition where the blood flow to the heart muscle (the myocardium) is occluded by a blood clot. This assertion is "accepted" in the sense that it is embedded in nutritional guidance issued by governments, though not in the sense that all scientists agree with it.

For the logical positivists, knowledge consists of propositions composed of verified units (standardized definitions of "saturated fats" and "myocardial infarction"), which are placed in logical relations ("consumption of saturated fats are a cause of myocardial infarction"). These "facts" are "verified" by empirical evidence (for example, data that show an association between the dietary intake of saturated fats and the incidence of myocardial infarction).

Myocardial infarction is defined as prolonged ischemia (restricted blood supply) leading to the death of myocardial cells.[7] Commonly, myocardial infarction is when a blood clot forms in an artery of the heart, and it can be diagnosed postmortem by evidence of blood clots in the arteries or chambers of the heart. In a living patient, it is diagnosed from symptoms such as pain or tightness in the neck, chest, and arms; or from abnormalities in electrocardiogram results; or, using angiography, by detecting blood clots in the coronary arteries.

Thus, myocardial infarction is defined partly by our theoretical understanding, and partly by operational definitions—diagnostic criteria. Studies of its incidence report only the presence of features that are (sometimes inconsistently) associated with a presumed underlying process. These might be symptoms that have proved useful for indicating that something has happened (e.g., tightness of the chest followed by a loss of consciousness), or features which fit our understanding of the events leading to myocardial infarction. But the diagnostic criteria have changed with technological advances and theoretical refinements, and even at the same time have been different in different places.

In 1968, a study on myocardial infarction by the UK Medical Research Council compared results obtained in London with those of a study conducted just two years before in Oslo. It found that "the Oslo criteria cannot be exactly applied to the London data."[8] The London criteria recognized seven cases of myocardial infarction in the test group and four in the control group in men under sixty years of age. By the Oslo criteria there were thirty cases in the test group and thirty-eight in the control group. Different ways of diagnosing myocardial infarction meant that two studies, although apparently studying the same phenomenon, had completely different results.

Thus, the "facts" of science do not always survive changes in conceptual frameworks; medical facts do not always even survive subtle changes in diagnostic criteria. Since 1968, the diagnostic criteria have been refined, particularly by the measurement of biomarkers—enzymes and other factors in blood samples that indicate myocardial damage. This has led to the diagnosis of myocardial infarction in cases with no other symptoms, and hence to an apparent increase in the incidence of myocardial infarction that purely reflects its redefinition.[9]

Saturated Fats

And what of "saturated fats"? Although this term is used in government dietary guidelines, it refers not to a single entity but to a class of fatty acids. Fatty acids are compounds composed of carbon atoms connected to hydrogen atoms. Saturated fatty acids (SFAs) have no double bonds between the carbon atoms—they are "saturated" with hydrogen atoms. Monounsaturated fatty acids (MUFAs) have a single double bond between carbon atoms, and polyunsaturated fatty acids (PUFAs) have more than one double bond.

According to the prevailing understanding, SFAs favor the build-up of fatty deposits in arterial walls, which, over time, become prone to rupture. These deposits are rich in cholesterol. Cholesterol is an important constituent of all cells in the body; it is produced in the liver, and transported to other sites in the body by low-density lipoproteins. When the blood levels of cholesterol are too high, it is believed to infiltrate the walls of arteries and provoke inflammation.

In the 1950s, scientists reported that different types of fat in the diet had different effects on blood levels of cholesterol and lipoproteins. At first, they classified fats simply as animal fats or vegetable fats. While animal fats appeared to raise total blood cholesterol, vegetable fats generally did not.[10]

By the end of the 1950s, studies had suggested that, whereas SFAs in the diet increased both total blood cholesterol and the amount carried by low-density lipoproteins, PUFAs lowered total cholesterol.[11] Thus, animal fats appeared to be more dangerous than vegetable fats because of their higher SFA content. Accordingly, it became common to distinguish between "saturated" fats and the presumably healthy "unsaturated fats"—PUFAs and MUFAs.

However, SFA is an umbrella term encompassing at least thirty-six known saturated fatty acids. Only some of them increase blood cholesterol. Lauric acid does, but short-chain SFAs have little or no effect. Bovine milk fat contains about four hundred different fatty acids.[12] The term "saturated fats" conceals a complex reality.

Nutrition scientists focused on the whole class of saturated fats because it was easier to measure the total saturated fat content of foods than their specific fatty-acid composition. When looking for associations between diet and heart disease in populations, establishing the specific fatty-acid composition was impractical and seemed unnecessary.[13] This made it easier to design epidemiological and intervention studies and helped policymakers to formulate a clear and simple message for the public. In the 1970s, US national dietary guidelines advised people to reduce their consumption of foods high in SFAs by avoiding dairy products and fatty meats.[14] They advised people to increase the consumption of PUFAs to about 10 percent of food energy by switching from fats such as butter to oils high in unsaturated fat, such as soybean oil.

Scientists used (and still use) the term "saturated fats" not because of its accuracy, but because it was easily understood. In shaping a simple account of the relationship between dietary fat and heart disease, scientists and policymakers relied mainly on evidence from observational studies of associations between diet and heart disease in populations, and evidence from intervention studies of the effects of particular diets on blood cholesterol. But, in 1973, Raymond Reiser, a specialist in lipid chemistry, raised concerns that the vagueness of the term "saturated fat" could mislead:

> A step away from accuracy (which is unacceptable) is the now common use of the term as being synonymous with animal fat. [. . .]. However, animal fats can be quite polyunsaturated if polyunsaturated fats are included in the animals' diets. [. . .] The misleading practice of referring to animal fats as "saturated fats" has arisen in epidemiological studies because only natural foods such as butter, cheese, and eggs are involved in population investigations.[15]

It is now clear that SFA is not the only type of dietary fat to raise blood cholesterol; indeed, it is not even the most dangerous. One category of unsaturated fatty acids is now believed to be more dangerous—*trans-fatty acids*.

Trans-fatty acids raise levels of cholesterol held in low-density lipoproteins, and reduce levels held in high-density lipoproteins. But it was these that industry replaced SFAs with, following what was understood to be the best scientific advice.[16] Fats high in SFAs are solid at room temperatures, whereas those high in unsaturated fatty acids form oils. For baked foods, solidity is important for texture and palatability, and one way to make unsaturated fats solid at room temperature is by partial hydrogenation—producing trans-fatty acids. Accordingly, food manufacturers reformulated popular foods by replacing butter with alternatives such as margarine that were presumed to be "heart-healthy" by virtue of their unsaturated fat content. But it seems that people who changed to these products did not protect themselves from heart disease. In 2005, the National Academy of Sciences stated clearly that "dietary trans-fatty acids are more deleterious with respect to coronary artery disease than saturated fatty acids."[17]

The case for the nutritional advice had come mainly from observational studies, with little understanding of the mechanisms involved. Such gaps in understanding still underlie doubts about the validity of nutritional advice generally. Any nutritional advice will have consequences; eliminating one food type from the diet will lead to its replacement with another. It will have unintended consequences, for good or ill.

In the case of such concepts as stress and intelligence, we saw fundamental problems in how to define them—the vagueness of the concepts reflects the uncertainty and incompleteness of our understanding. The term "myocardial infarction" seems to have a precise definition—but how it is defined differs over time and from place to place. The term "saturated fat," by contrast, has a clear, precise and stable definition, yet it is used loosely, as though all saturated fatty acids have properties that only some do.

The Sokal Hoax

When vagueness becomes incoherence perhaps it deserves to lose all respect, as illuminated entertainingly, if unfairly, by the Sokal Hoax.

In 1994, a physicist, Alan Sokal, frustrated by the apparently impenetrable prose of certain social constructivists, submitted a nonsense paper entitled "Transgressing the Boundaries: Towards a Transformative Hermeneutics of Quantum Gravity" to one of their academic journals, *Social Text*. The paper was sprinkled with quotations from prominent constructivists and might be seen as a test of whether even social constructivists could recognize that they made no coherent sense, at least in the context in which they were quoted. It was accepted for publication, and the hoax once exposed became famous—or infamous.[18]

Social constructivism seeks to understand knowledge as the product of social interactions. It recognizes the importance of societal interests: how they determine the resources devoted to particular avenues of science and how this influences the development of scientific knowledge. It recognizes the importance of historical paths in how theories develop. It recognizes that ideological and commercial influences affect what, at any given time, is understood to be knowledge. It recognizes that, at different times and in different circumstances, different standards of evidence have been applied, and that these are determined by social conventions observed within communities of scientists. It examines how vested institutional interests influence scientific approaches to class, gender, race, and sexual preference.

A common conclusion of social constructivists is that, rather than being driven by a rational scientific method, science takes the form that it does because of the way that these social processes work. In an extreme form, these considerations have led some to the apparently absurd conclusion that "the natural world has a small or nonexistent role in the construction of scientific knowledge."[19]

What the hoax actually tells us is questionable. It was conceived as a challenge to the notion that facts and evidence merely reflect subjective interests and perspectives. What it seemed to illustrate is that this notion was indeed held by editors and referees of *Social Text*. They did not concern themselves with questioning the facts or evidence in the hoax paper, but were apparently impressed by the support for the totems of their field, regardless of whether that support was expressed appropriately or not.

Perhaps the editors took seriously the idea that science should be "democratized"—that any notion generated with apparent serious intent should be published, that "experts" should not be censors and gatekeepers of conventional wisdom, and that readers should be left to decide for themselves what is nonsense and what is not. This is an opinion held by some scientists who have become frustrated by the vagaries of peer review.[20] By this view, the only role of a journal is to select contributions that its readers might be interested in.

We cannot draw robust conclusions from a single case study, however entertaining. Some bad papers are accepted by all journals, and papers in all fields are more likely to be accepted if they seem to conform to accepted notions. Impenetrable writing often goes unquestioned. Many referees, when faced with dense, portentious prose from an established authority, seem to accept that their inability to understand it reflects their own intellectual weakness rather than that of the authors.

Quotations out of context can often be made to appear nonsensical. But sometimes they *are* nonsensical, and even where they are not, there can be no robust defense of making sensible statements incomprehensible. The virtue of clear writing is that flaws in reasoning can be more readily appreciated. To do so requires separating the logic from the rhetoric.

Consider this example from Sokal's attack on a paper by Barry Barnes that sought to explain the sense in which sociologists understand scientific truth to be relative. Barnes had written "There is a certain very obvious sense in which different *maps of a given terrain* should be understood relativistically, as all on a par with each other" (emphasis added).[21]

Sokal quoted this passage in full, and in his attack, he acknowledges that different maps—for instance a road map and a contour map—depict different aspects of reality. But he goes on to state that "to say that no map stands in closer correspondence with the mapped terrain than any other is absurd. Would Barnes really want to spend his holiday in Paris navigating with a street map of New York?"[22]

Before we celebrate too vigorously the scientist's put-down of a sociologist, we should ask in what sense are street maps of New York and Paris both "maps of a given terrain." They are not. They are maps of different terrains. Nor would any reasonable reader of Barnes imagine that he is arguing that all maps are equivalently good for any given purpose, his point is that the value of each is relative to its purpose. The map of the London Underground is a design icon for its clear and efficient rendering of a complex transport system, but anyone who tries to use it as a guide to walking through London will soon find that the spacing of the stations bears little relation to their physical proximity.[23]

The London Underground map, the London tourist map, and the Ordnance Survey map of London serve quite different purposes, and their purposes define their forms. There is no sense in which one map is better than another, except by reference to the interests of their users. Each map uses data selectively, and uses it to enhance a communicative intent. None of the maps is "stable," each needs regular updating, and none is a unique solution even for the problem that it engages with.

A complex scientific statement is like a map of the London Underground. That map has a defined connection with the real world—it helps us to travel from one place to another. It is clear, and to this end it simplifies and neglects much detail. It is not the only possible map of the Underground—others might display the physical location of stations better, but in doing so may be less understandable. It is the way it is because it suits our cognitive capacities and because it encourages more people to use the Underground.

The Underground Map has an immediate purpose, to direct travelers efficiently to their chosen destination. By contrast, the immediate purpose of any scientific paper is to direct readers to a destination chosen by someone else—by the author. It aims, at least in part, to persuade its readers of the validity of the ideas or of the findings that it describes, and its intended readers are those who may see potential utility in those ideas and findings. It's a purpose that would be rendered futile if the conclusions of that paper could be readily contradicted by reason or by experience. But it's a purpose that is also supported by rhetoric, selective citation, and selective presentation of evidence.

It seems legitimate to inquire whether the ideas expressed in papers are influenced, for example, by ideology, and whether how those ideas are received is influenced by the social status, sex, or race of the authors. It also seems legitimate to inquire whether findings are influenced by commercial or other interests of funders, and to consider whether the personal interests of scientists in advancing their careers might influence how they present their findings.

It seems not just legitimate, but important, to know whether particular "facts" are accepted on the basis of the weight of evidence, or for their utility, or for other reasons.

9 The Totality of Evidence: Weighing Different Types of Evidence

> A verdict will be given in favour of the side bearing the legal burden of proof only if, having considered all of the evidence, the fact-finder is satisfied that the applicable standard of proof is met."
>
> —Hock Lai Ho[1]

Scientists differ about how evidence should be interpreted, what methods are appropriate for testing a hypothesis, whether a phenomenon has been measured or categorized correctly, and whether the hypothesis is properly grounded in other evidence and theory. This poses problems for those who seek to weigh the totality of evidence. Yet this must be done if we aspire to evidence-based policies on any issue, from environmental policy, to clinical guidelines, to public health policy.

Type 2 diabetes is a disease of enormous concern; it affects about 400 million people worldwide and incurs huge costs through health care expenses and time missed at work, in addition to high social costs. Most cases are thought to have environmental causes—and so changes in lifestyle might prevent them. Clearly massive benefits might be achieved by changing lifestyles in the population to reduce risk. But when is there enough evidence to change public health policy?

Fish, Pollution, and Diabetes

Suppose that an association is found between fish and the incidence of type 2 diabetes. What should we make of this? We might think of many possible explanations:

- eating fish can cause diabetes;
- diabetes causes people to eat more fish;
- people who eat more fish eat less of some other things that can protect against diabetes, or change their behaviors in other ways that promotes diabetes;
- the consumption of fish and the incidence of diabetes are both independently affected by some third, unknown factor;
- the apparent correlation is a "false positive" outcome of the statistical tests used; or
- the apparent correlation is the consequence of some technical error.

The mere observation of an association gives the scientist no way forward. It is only when she relates it to a whole body of background information, theories, hypotheses, and assumptions that the prospect of any path will appear. In this case, she is likely to pursue the hypothesis that eating fish can cause diabetes— not because it is the most likely, but because it is the only one of the various hypotheses that seems testable and which, if true, would be interesting and possibly important.

Testing this hypothesis is not easy because "causes" in this context are not strictly deterministic. In this kind of business, a significant association might be merely that, all other things being equal, people who eat a lot of fish are on average 20 percent more likely to get diabetes than other people. In public health terms, such an association, if upheld, would be one that is very strong, certainly strong enough to merit contemplating some action.

If the prevalence of diabetes is about 6 percent in the adult population, as it is in the United Kingdom, a 20 percent increase in risk means that for every one hundred people that eat a lot of fish, there will be about one more case of diabetes than would otherwise be expected. To verify the original observation we might start by trying to replicate it, but to do so robustly is no trivial undertaking. You can look at a lot of people and ask them about their diets and hope that they don't lie or misremember, but that little phrase "all other things being equal" is a nuisance.

The incidence of diabetes varies with age, body weight, gender, ethnicity, exercise levels, and other features of the diet— these are things that we know, and these we must consider. We will have to look for an effect of eating fish that goes beyond the effects that can be attributed to how these other factors differ between fish eaters and fish skippers. However, we can only allow for factors that we know about and can measure. Diabetes has a strong heritable component, but we don't know (yet) all of the genes involved. We probably can't afford to sequence the genomes of all of the subjects of such a study and wouldn't know enough about what that tells us anyway, so we just can't consider this factor.

So how do we proceed?

Three Pillars of Evidence

Generally, evidence of a health risk requires three "pillars"— evidence of association, evidence from intervention studies, and evidence of mechanism. Evidence of association is often considered to be the weakest of these, because no evidence of an association can legitimately be interpreted, in itself, as a demonstration of causality. Nevertheless, evidence of association is the usual starting point, perhaps something noticed in a particular community, perhaps an observation incidental to the primary purpose of some study. Such reports are often false positives, but sometimes more scientists will start to look for evidence, and sometimes a case will develop that seems to need deeper investigation. Even if there is only erratic evidence of association this is not necessarily the end of the story. We might find a strong association in some communities but not in others—this would raise the question of whether other factors, perhaps in combination with fish consumption, might be involved.

Evidence from intervention studies is clearly to be desired—can we prevent by limiting the consumption of fish, or provoke it by replacing meat with fish in the diets of fish skippers? The latter we wouldn't contemplate if we think that fish might add to the risk of diabetes, and the former is no less problematic, even if we can get volunteers to comply with a study, there are other acknowledged benefits of eating fish that it would be reckless to forego.

We might look for interventions in an animal model. There is extensive conservation of biological mechanisms between species: these arise by our evolution from common ancestors. That theory, one of Charles Darwin's great contributions, is often linked with, but is logically independent of, the theory of evolution by natural selection. Both theories were maintained not on the merits of Darwin's evidence and reasoning alone, but first because they could be united with Gregor Mendel's theories of heredity in a joint framework, known as the "Modern Synthesis," and second because they could be integrated with the understanding of genes that developed through molecular biology. With this, the theory of evolution from common ancestry, once thought to be questionably scientific because of the inability to test it, came into the scope of potential falsification. The theory now had predictive potency—the genes of all animals should display, in their structure, evidence of common ancestry. With the ability to sequence whole genomes quickly and cheaply, that unity became evident.

Although there is deep conservation of genes, there is also extensive diversification of form and function—our diets are unlike those of most other animals, and our guts differ by their evolutionary adaptation to our different diets. We can use animal models, but not crudely; we must use them to study things that have been conserved through evolution, not things that have diverged. We cannot sensibly try feeding fish to mice.

So, here, as often, the key pillar is evidence of mechanism. We must ask what mechanisms might account for a causal link between eating fish and diabetes?

To make progress on this requires another host of diverse theories and hypotheses and a vast background of knowledge and understanding about other things, most of which are apparently unrelated to the issue at hand. We must consider what we already know about diabetes—that it affects cells in the pancreas that produce insulin by killing them or otherwise preventing their normal behavior. We might recall that we know of other things that can cause diabetes, including genetic causes, obesity, and various toxins.

Amongst the toxins that can damage these cells are some environmental pollutants. Polychlorinated biphenyls (PCBs) are chemicals that once had widespread industrial use, but whose production is now banned because of their toxicity. PCBs do not easily degrade—this made them attractive for industrial applications, but meant that PCBs can enter the environment and the food chain. They entered the environment through paint, hydraulic fluids, sealants, and inks disposed of in waterways and dumps from which they entered rivers and groundwater, ultimately reaching the sea. They are poorly soluble in water but very soluble in oils, and they accumulate in the mud of rivers and marine environments. From here, they are absorbed by shellfish and other animals that dwell in the mud. They accumulate in the bodies of fish that eat those animals, and at still higher concentrations in the bodies of animals that eat the fish—including people. They enter the food chain, and the animals in that chain differ in their ability to metabolize and excrete them.

Because PCBs are soluble in fats, they accumulate in mammals in the cells that store fat, the adipocytes. The insulin-producing cells in the pancreas are surrounded by adipocytes and have an intimate relationship with them. Insulin is a regulator of energy availability in the body, and its secretion is modulated by the amount of energy reserves held in fat tissues.

One theory of the causes of diabetes is the *lipid toxicity theory*— that when the amount of fat in the body exceeds the capacity of adipocytes to hold it safely, excess lipid spills over into adjacent tissues and into the blood with toxic effects. These might be either the direct effects of excess exposure to lipids, or indirect effects from other chemicals dissolved in those lipids.

In this case, progress in testing our hypothesis (that eating fish can cause diabetes) is likely to depend on many things. It requires understanding the biology of the insulin-producing cells, how PCBs affect them, how PCBs are metabolized in all animals in the food chain, what other consequences arise from exposure to PCBs, what other risk factors might be involved, and what other consequences might be expected from changing the diet to exclude fish. At the same time, we must consider alternative hypotheses. Perhaps the problem lies with the gut microbiota. Our gut supports a massive diversity of microorganisms that are important in digestion, and the species present in the gut depend on our diets. Not all of these species are benign; some are associated with inflammatory conditions, and type 2 diabetes is one such condition. Perhaps it's not all fish that are problematic, but only oily fish with a high fat content—but these contribute nutritional benefits as a source of long-chain omega-3 fatty acids. There is an inevitable "path dependency" in the pursuit of any understanding. As a research program develops, scientists become committed to a single hypothesis because they have invested heavily in it—not always because they have systematically refuted the alternatives.

This is not a hypothetical example. Globally, type 2 diabetes has become increasingly prevalent, but especially among indigenous populations. In Canada, its prevalence among aboriginal peoples of First Nations heritage is 3–5 times higher than in the general population leading to higher mortality. There are many genetic and lifestyle risk factors, but environmental contaminants may also contribute to these disparities. Epidemiological studies have found that, in some communities, type 2 diabetes is positively associated with exposure to PCBs; in Canada, First Nations people are exposed to higher levels of PCBs than the general population, apparently through fish consumption.[2]

Triangulation

The research strategy alluded to earlier is sometimes called *triangulation*. The principle is that when diverse sources of evidence converge on the same conclusion, then our confidence in that conclusion is increased. Here, the elements of the triangulation are three types of evidence: evidence of association, evidence from intervention studies, and evidence of mechanism. Each of these has its flaws.

If there is no evidence of an association between observed factors and an outcome, then it seems reasonable to conclude that there is no important causal link between them. But no amount of evidence of association alone can reliably lead to a conclusion about causality.

Intervention studies are problematic for different reasons. Typically, they aspire to being one-dimensional—to vary one experimental factor while holding all else constant. In practice, changing one factor always results in changes to others— sometimes in unforeseen ways, sometimes in ways not noticed, sometimes in ways constrained unrealistically by the design of the study—and one factor might have effects only in combination with another.

Studies of mechanism generally involve a particular experimental setting—they might for example involve model organisms or cell lines maintained in vitro. Individual mechanistic experiments are typically one dimensional in design, but a mechanistic research program is likely to involve a wide variety of experiments targeting each link in a hypothesized causal chain. This entails another source of vulnerability—a chain is only as strong as its weakest link.

Thus, the type of science needed to address an apparently simple question—such as "Can eating fish cause diabetes?"—is dauntingly complex. It should not be a cause for surprise that, nearly a hundred years after the first observation that a high-fat diet might be linked to heart disease, the nature of that link and even its existence in human populations is still a source of dispute.

In recent years, the European Union has funded large, multinational, multidisciplinary consortia to tackle such questions—problems seen as requiring a diversity of expertise that is beyond the resources of individual nations. These projects have had unexpected benefits. When the collaboration is meaningful, it forces scientists to express their reasoning again in unfamiliar terms—they must convince others, others who may not always be impressed by the arguments that satisfied their immediate peers. This forces them to come to terms with different bodies of theory and accepted knowledge—and to grapple with the contradictions.

Yet there are also problems. When multidisciplinary projects are proposed to funding agencies, and when the outcomes of multidisciplinary collaborations are published, the peer review process is weaker: few reviewers may have the breadth of technical understanding to judge the projects or their outcomes in totality. Moreover, when more people and more resources are involved in a collective program, the collective investment is greater, and the pressures on individuals to produce results that advance the common cause is greater. The shared bonds of expertise are weaker, and the reliance on perceived authority is greater.

The big picture can hide ugly detail.

Underdetermination

Thus, the question "Can eating fish cause diabetes?" is not a simple one, and it engages one of the most influential ideas in the philosophy of science: the Duhem–Quine thesis of underdetermination of theory by empirical evidence.[3] Pierre Duhem (1861–1916) was a theoretical physicist who argued that no experiment, however well designed, tests a single hypothesis in isolation. It is always a test of the entire interlocking body of assumptions, theories, and hypotheses: "when the experiment is in disagreement with his predictions, what he learns is that at least one of the hypotheses constituting this group is unacceptable and ought to be modified; but the experiment does not designate which one should be changed."[4]

The philosopher Willard Van Orman Quine (1908–2000) took this further; he argued that *all* of our knowledge and beliefs "is a man-made fabric which impinges on experience only along the edges. [. . .] the total field is so underdetermined by its boundary conditions, experience, that there is much latitude of choice as to what statements to reevaluate in the light of any single contrary experience."[5]

The Duhem–Quine thesis thus holds that the evidence available is, in itself, always insufficient to justify our belief in our theoretical explanations. Abstract reasoning from evidence cannot therefore be the decisive factor in selecting between theories that can explain the evidence equally well.

Popper's answer to this challenge was that scientists choose theories and hypotheses not only because they can explain the evidence, but also because they are simple, bold, and testable. By "bold," he meant that had extensive predictive potency, and by "testable," he meant that they could be tested against their predictions. But "simple" in this context seems to have no objective meaning. What a mathematician understands by "simple" can be very different from what an experimental scientist would understand by it. Galileo's conception of a heliocentric solar system was, as correctly observed by Pope Urban VIII, a mathematical representation of the solar system that was logically equivalent to an alternative representation that placed the Earth as the center of the system. Accordingly, science and mathematics alone could not be the final arbiter of objective truth. The Pope held that the Church was properly the final arbiter; Galileo held *a priori* that the laws of nature were mathematical in form, and hence the simplest mathematical representation was "true."

Scientists do not struggle with such ideas: their choice of a hypothesis is pragmatic and utilitarian. They select a hypothesis that is 1) simple, in being easy to explain (to the relevant audience); 2) open to testing by means currently readily available; 3) interesting, in having implications of practical value. By these criteria, scientists are not faced with a multitude of hypotheses to choose from. As one physicist put it, "The problem, more often than not, is to find even a single reasonable theoretical structure that can comfortably embrace a large and expanding body of accepted knowledge."[6]

Finally, the choice of a hypothesis is path-dependent. A hypothesis emerges from a previous hypothesis, and in response to a specific challenge. Once a hypothesis has been conceived it is sustained because scientists invest in it. They continue to do so until the returns on their investment have diminished to the point at which it becomes worth shifting to an alternative hypothesis, despite the temporary losses that that may entail.

10 Exaggerated Claims, Semantic Flexibility, and Nonsense

> When anyone tells me that he saw a dead man restored to life, I immediately consider with myself whether it be more probable that this person should either deceive or be deceived or that the fact which he relates should really have happened. I weigh the one miracle against the other and according to the superiority which I discover, I pronounce my decision. Always I reject the greater miracle. If the falsehood of his testimony would be more miraculous than the event which he relates, then and not till then, can he pretend to command my belief or opinion.
>
> —David Hume[1]

In chapter 6, we saw that when a position is advanced in science that poses an existential threat to a community of scientists, when it questions the value of their work and undermines their credibility, the response of those scientists is likely to be acerbic. Scientific discourse is normally polite, at least as seen from outside by observers oblivious to the acid hidden in seemingly bland phrases. There are some circumstances, however, in which vitriol is undisguised. When scientists seem to promote claims that breach reason by their disregard of prior knowledge, the response of other scientists can be brutal. So too when scientists seek to elevate the status of "claims" to the status of "facts" without "due process" in interpretation—without relating them adequately to theory, and thereby to a broader body of knowledge.

Genomes

When the Human Genome Project was launched in 1990 to determine the sequence of the 3 billion chemical units (the "base pairs") that comprises human DNA, molecular biologists had little idea about how many protein-coding genes would be present in this sequence. In 2000, Ewan Birney, later to be codirector of the European Bioinformatics Institute, took the first bets on the number of genes at a bar during a genetics conference, launching a contest that attracted more than a thousand entries.[2] The guesses ranged from about 26,000 to over 312,000. The final tally came in at about 20,000; the human genome has more genes than that of the chicken, but far fewer than that of the onion.[3]

Only about 1.5 percent of the human genome consists of protein-coding genes, so what is the rest for? It seemed that large amounts of DNA had to be junk. Two arguments underpinned this conclusion. First, even some closely related species had genomes of massively different sizes—the size of the genome in a species seemed to bear no relation at all to its place on the evolutionary tree. The genome of the Mexican axolotl (a species of salamander) is ten times larger than the human genome, while that of the Japanese tiger pufferfish is about one seventh as big. Second, the size of an organism's genome bears no relationship to its complexity: one of the largest sequenced genomes is that of a single-celled animal, *Amoeba dubia*; with 670 billion units of DNA, its genome is two hundred times larger than the human genome.[4]

Junk DNA

The term "junk DNA" was introduced into the scientific lexicon in 1972 by the geneticist Susumu Ohno (1928–2000).[5] Ohno had studied the rate at which random, spontaneous mutations arise in the genome in each generation. Such mutations matter if they occur in a *gene locus*—in the gene itself or in its "promotor" or "operator" regions (regions of DNA, usually close to the gene that determine how the gene is regulated, including in what cells it will be expressed and in what quantities). A few mutations will be beneficial, some will be neutral, but many will be deleterious. He estimated that, at any given gene locus, in any single generation, there is a 1 in 100,000 probability of a deleterious mutation occurring there. Deleterious mutations are eliminated by natural selection—but the rate at which this can occur is limited; it is a function of selection pressure, and even when the selection pressure is strong it can take many generations to eliminate a detrimental mutation from a population. Ohno thus saw that if detrimental mutations accumulate faster than they can be eliminated, there would be a progressive decline in genetic fitness. He estimated that only about 6 percent of the human genome could be potential targets for deleterious mutations—that is, could be functional in the sense that damaging them had a deleterious effect. From the size of known genes, he estimated that the human genome must contain only about 30,000 gene loci. Ohno's estimate came about thirty years before the actual number of genes was shown to be about this.

Knowing the rate of spontaneous mutation was key to understanding how new genes arise. Calculations show that it is inconceivable that a new gene can arise through random, spontaneous mutations at discrete points in the genome. Instead, new genes generally arise because of mutations that involve the duplication of an entire gene. When this occurs, the new copy is "redundant," and further random mutations in this copy will accumulate without selection pressure against them. Occasionally, such mutations will lead to new functions for this copy, and a "new" gene is born. But for every gene born in this way, there will be many "failed experiments" of nature. As Ohno put it, "The earth is strewn with the fossil remains of extinct species; is it a wonder that our genome too is filled with the remains of extinct genes?"[6]

In addition to pseudogenes, about 90 percent of the genome is littered with repetitive, mutationally degraded material with no apparent function. About 10 percent of the human genome consists of about 1 million copies of just one of these "parasitic" sequences, a sequence of about three hundred bases; these "Alu elements" have been implicated in many hereditary diseases, including hemophilia and breast cancer. Such elements have been called "parasitic" or "selfish" because of their ability to multiply while serving no useful function.

The ENCODE Consortium

Nevertheless, some of the genome contains regulatory regions that control where in the body and in what amounts the products of genes will be expressed. It also contains many genes that do not make proteins but that do make RNAs that have a role in protein synthesis. The DNA segments that have such biological "functions" tend to display certain properties. For example, proteins that regulate gene expression usually bind to certain specific sequences of DNA. It was thus proposed that a search for segments of DNA that had these properties would discover all the "master controllers" that controlled the genes.

Accordingly, a massive collaborative project was launched to produce an Encyclopedia of DNA Elements (ENCODE) with the aim of building "a comprehensive parts list of functional elements in the human genome, including elements that act at the protein and RNA levels, and regulatory elements that control cells and circumstances in which a gene is active."[7] It began in 2003 with a pilot project that aimed to survey 1 percent of the human genome, and in 2007 was expanded to a survey of the entire human genome and later to encompass other genomes.

The project produced an extraordinary amount of data that, according to one commentator in 2016, "represents a major leap from merely describing and comparing genomic sequences to surveying them for direct indicators of function," data that "can serve as a map to locate specific landmarks, guide hypothesis generation, and lead us to principles and mechanisms underlying genome biology."[8]

In 2012 the ENCODE consortium, accompanied by an orchestrated public relations campaign, appeared to claim that 80 percent of the human genome serves some purpose. This claim was rapidly picked up by science journalists who described the results of the project as heralding "the death of junk DNA."[9]

The Immortality of Television Sets

The apparent conclusion that 80 percent of the human genome is functional, flew in the face of accepted understanding. Criticisms were swift to follow, and Dan Graur's assault in 2013 on this idea makes Trewavas's prose (see chapter 6) look anemic. In a paper entitled "On the immortality of television sets: 'function' in the human genome according to the evolution-free gospel of ENCODE," he and his colleagues argued that the consortium had "fallen trap to the genomic equivalent of the human propensity to see meaningful patterns in random data." They argued that an "absurd conclusion" had been reached through a series of logical and methodological transgressions. They noted that the ENCODE results "were predicted by one of its authors to necessitate the rewriting of textbooks. We agree, many textbooks dealing with marketing, mass-media hype, and public relations may well have to be rewritten."[10]

Graur argued that the ENCODE authors had engaged in a logical fallacy known as affirming the consequent: from finding features in regions of the genome that were possessed by regions known to have a biological function, they had illegitimately concluded that all such regions were also functional. The argument in part was about the use of the word "function." In natural language, the "function" of something is the purpose that it serves, or an action for which it is specifically designed. For a biochemist, if a DNA sequence binds a protein then it is functional in that it does something. But this does not imply any physiological consequences for the organism. There is no purpose that is implied, nor is there necessarily any specific design involved.

Graur's paper was characterized by one commentator, Sean Eddy, as "angry, dogmatic, scattershot, sometimes inaccurate." It launched an "undignified academic squabble": "Attention focused on the squabbling more than the substance, and probably led some to wonder whether the arguments were just quibbling over the word 'function.'" But Graur's attacks hit home. In Eddy's words again, "It is bewildering to see ENCODE take an eminently reasonable data generation project and spin it so inexpertly as a hypothesis test with supposedly revolutionary conclusions."[11]

At the heart of the argument was a semantic issue: how the word "function" was used. The ENCODE consortium had appropriated this word to express a technical meaning specific to a narrow context, but had then presented their results as implying "functional significance" in a much wider sense. This a common problem—as we observed in chapter 8, many terms such as "stress" and "hunger" are used by scientists in a variety of technical ways that are all different from the way that they are used in natural language. Scientists are under pressure to explain their work to the public, so it is unsurprising that their messages are sometimes lost in translation, but it is particularly unfortunate when scientists confuse themselves. It is a common complaint that scientists use too much jargon in what they write and say, but sometimes they don't use enough.

Unbiased Research

But the arguments over ENCODE were also fueled by unease over "big science" approaches in biology. There is a movement against hypothesis-driven science that goes sometimes under the tag "unbiased research." This includes "high-throughput" research, such as the ENCODE project, that involves a comprehensive acquisition of diverse data which are analyzed to identify associations that might lead to novel insights. In biology, such approaches have been driven by new technologies. *Genomics* allows the sequencing of the whole genomes of individuals. *Proteomics* identifies the entire complement of proteins produced by an organism. *Transcriptomics* identifies the set of RNA molecules expressed by one cell or a population of cells. *Metabolomics* identifies the set of metabolites that arise from chemical processes within cells or within an organism. *Connectomics* maps the totality of connections between nerve cells in the brain. Older approaches that harness vast amounts of data include *epidemiology*, a foundation of public health policy which seeks to infer the causes of disease from associations between disease and environmental factors, and there are many more.

The scale of these activities and the speed at which they are expanding are startling. The Human Genome Project, initiated in 1990, included the aim of determining the DNA sequence of the human genome—about 3 billion base pairs—in just fifteen years. The project, a multinational collaboration including partners from eighteen countries was declared complete, two years ahead of time, in 2003. The actual cost of sequencing the genome had been between $500 million and $1 billion.[12] By 2016, new machines could sequence a human genome in about an hour for about $1,500, and the cost is still falling.

In 2012, the UK government launched the 100,000 Genomes Project to sequence the genomes of patients with rare diseases, some types of cancer, and infectious diseases.[13] Such approaches follow the belief that if you collect enough data from as many observables as possible, then by informatics approaches it will be possible to recognize chains of causality amongst them. Thus, some have claimed, you do not need a particular hypothesis: the data will tell their own story. By this view, hypothesis-driven science is intrinsically biased—if you pursue a hypothesis, you will only find what you look for, and will be blind to interpretations other than those that you are already primed to expect.

High-throughput research is, sometimes grudgingly, accepted as a necessary evil. It is expensive and diverts funding from traditional hypothesis-driven research. The apparent notion that if enough data are collected, they will tell their own story, is often greeted with derision. A more refined view is that it is never possible to deduce causality from associations alone, but finding an association is a way of inspiring a new hypothesis, which must then be subject to test. Any nontrivial hypothesis requires *some* foundations—piles, however strong, do not stand without *any* support—and a bold hypothesis is formulated most readily where there is enough knowledge to generate questions.

> **Pareidolia:** the tendency to perceive a specific, often meaningful image in a random or ambiguous visual pattern.[14]

A problem with such "big science"—but also with much "little science"—is our human propensity to recognize shapes in clouds, to hear messages in noise, to see patterns in random arrays of dots or numbers. In 1955, the psychologist Michel Gauquelin, in *L'influence des astres* ("The Influence of the Stars"), reported that a significant number of sports champions were born just after the planet Mars rises, or when it crosses the meridian, as would be observed at the place of his or her birth.[15]

Given the notice that this book received, the US Committee for Skeptical Inquiry and the Comité pour l'Étude des Phénomènes Paranormaux (Committee for the Study of Paranormal Phenomena) in France undertook large replication studies, and both failed to reproduce the correlation.[16] These studies recognized several flaws in Gauquelin's work. The major flaw is one common in "mainstream" scientific papers—the flaw of not adequately accounting for multiple comparisons.

The problem is well known by most scientists, though sometimes forgotten in practice. The outcome of every statistical test is subject to chance—it might be true, but it might be either a "false positive" or a "false negative"—and, conventionally, scientists use an arbitrary threshold criterion to determine what they present as statistically significant. That threshold is commonly set at $P < 0.05$. The P value has a technical meaning: it is the probability of finding a difference at least as great as the observed difference by chance alone—that is, of finding such a difference between random samples from a single population. Informally, $P < 0.05$ is taken to mean that there is less than a 5 percent chance that the result of the test is a "false-positive." If you perform fourteen such tests on random data, it is more likely than not that one of them will give a false positive result.

The problem of multiple comparisons is that the more tests you apply, the greater the chance that one will return a false-positive result. Accordingly, in rigorous statistical analyses, the P value must be corrected to account for this—the test becomes less powerful as a consequence and compensating for this requires a larger sample size.

However, this leads to issues with "exploratory studies"—when many factors are measured in a study that seeks to explore possible associations between them. If you collect data on just ten factors, to explore all associations between any two of them will involve more than 18,000 comparisons. To detect any true association in this way will require a very large sample indeed.

Nobody does 18,000 statistical tests in this way. They do just a few, perhaps between factors that are suspected to be associated based on other evidence. But if so, then decisions about what to compare could have been made before the experiment. Today, clinical trials in medicine are expected to follow robust methodological guidelines. In a well-designed trial, the choices of what to compare and how to compare them are chosen beforehand and are declared in the process of preregistering the trial. This designates "primary outcomes" of the study, and this prior declaration of the hypothesis permits the use of powerful tests—the exact nature of which must also be registered.

Few experimental studies are preregistered, however, and most multifactorial studies are exploratory, in that the researchers suspect that some associations will exist but are unsure exactly where. In this case, the researchers are likely to choose which factors to compare based on the outcome of their studies—they will perform tests only between factors that appear to be associated. This is like estimating the weight of people in different cities by sampling their populations, and comparing only the heaviest group—in, say, Edinburgh—with the lightest group—in, say, Belfast—to conclude that Edinburgh people are significantly fatter than Belfast people. If this ignores the samples, in London, Cardiff, Bristol, Dublin, and elsewhere, then the test is illegitimate and the conclusion invalid. If the eventual paper reports only the data from Edinburgh and Belfast, these missing multiple comparisons are invisible.

Alternatively, the researchers might retrospectively construct a hypothesis and claim that the intended primary outcome was a comparison between Edinburgh and Belfast based, for example, on known differences in their diets. They might thus present their whole data, singling out the Edinburgh–Belfast comparison as the primary focus of their study and showing the remaining data as exploratory findings, for information only but not subject to test.

The problem of hidden multiple comparisons was exposed by Karl Peace and colleagues in a study of meta-analyses in nutritional epidemiology.[17] They looked at meta-analyses of the association between the consumption of sugar-sweetened beverages and type 2 diabetes. Many of the primary studies that were used in the meta-analyses had made no allowance for multiple comparisons. Peace et al. looked at each of the ten papers used in one meta-analysis and counted the numbers of outcomes, predictors, and covariates to estimate the size of the analytical "search space"—that is, the total number of possible comparisons. From reading the abstracts of the papers, which report the key findings, it appeared that the median number of possible comparisons was just 6.5. But the full texts showed a median number of possible comparisons of 196,608; each study had between 60 and 165 foods that could be predictors. Peace et al. concluded that "the claims in these papers are not statistically supported and hence are unreliable so the meta-analysis paper is also unreliable."[18]

This is a problem in many areas of science, including when apparently objective outcomes derive from state-of-the-art technologies. In 2018, Sean David and colleagues analyzed 179 studies using functional magnetic resonance imaging (fMRI)—"brain scans"—that had reported sex differences in the size of various brain regions.[19] Such differences are to be expected: as Cordelia Fine noted, a few brain regions are strikingly sexually dimorphic, and some small differences are likely to be present in many areas.[20] In some cases a biological interpretation can be attached to them—some brain areas regulate sex-specific reproductive functions, for example, and some of these areas differ in ways that are developmentally determined. But, as the brain is malleable and develops throughout life according to experience, some differences will simply register the consequences of cultural or lifestyle differences and have no implications for cognitive capacity. Fine concluded that there was widespread publishing of small, underpowered fMRI studies that propagated false-positive claims of sex differences in the brain and that these had led to "the proliferation of untested, stereotype-consistent functional interpretations."

Most of the studies reviewed by David et al. had small sample sizes. A study with a small sample is likely to miss small differences because it is likely to fail to reach the threshold of statistical significance. As small fMRI studies are common, many should find no significant sex differences, even if genuine differences exist. However, of the 179 studies, 158 reported finding one or more significant differences in some region or another. This was far more than expected from the sample sizes and the size of the differences.

This seems likely to reflect two biases in the literature: *publication bias*, whereby studies that find no significant differences go unpublished, and *selective outcome and analysis reporting bias*. David et al. concluded that these biases stem from "large flexibility in the modes of analyses, inappropriate statistical methods, and selection pressure from the current reward and incentives system to report the most significant results."[21]

Detecting such problems presents challenges to the referees of a paper. They might suspect that more things were measured than have been reported, that implicit multiple comparisons have been ignored, or that the hypothesis as presented is a post hoc artifice. But these things are generally impossible to be sure of. There is no reason to think that such flaws are always conceived deliberately; generally, scientists will convince themselves that they had a sound rationale for analyzing the data the way they did, even if they had not realized it until they saw the outcomes of their study.

Back to Astrology

Gauquelin's astrological findings were taken seriously by Barry Barnes, David Bloor, and John Henry, leading scholars of the sociology of science.

> Michel Gauquelin's statistical evidence in support of astrology would perhaps be a serious embarrassment to scientists if they were not so good at ignoring it. But one day it could conceivably come to be

accommodated as a triumph of the scientific method. Gauquelin's work seems to imply the existence of forces and interactions unrecognized by current scientific theory and yet it is based on methodological principles and empirical evidence which have so far stood up to sceptical challenge.[22]

These authors can be forgiven for being unaware of the statistical weaknesses of Gauquelin's evidence. In many fields, most scientists have little insight into even the tests that they themselves apply, and flaws are common in even the supposedly best journals. Steve Goodman, a medical statistician and editor of the *Annals of Internal Medicine* has been quoted as saying that he rarely sees papers that are free of flaws: "I view most of the literature as done wrong."[23]

Most scientists ignored Gauquelin's work; scientists ignore vast swathes of papers in areas where they have no interest or expertise, and at least provisionally, they ignore many of the papers in their own field that do not reach a preliminary bar of credibility. While there is a poor understanding of the nature of the many common flaws in statistical analyses, their prevalence *is* widely recognized as part of the "noise" in the literature. Evidence of association is, at best, weak evidence of causality, and it only gains credibility when conjoined with other forms of evidence, such as evidence from prospective studies, or from intervention studies and, critically, evidence of mechanism.

Gauquelin's paper offers scientists no way forward, however open-minded they might wish to be. It is hard to conceive of an intervention study without a means of disrupting Mars's orbit, and evidence of mechanism demands a sophisticated framework of theory integrating diverse sources of evidence—a framework wholly absent in this case.

A very few scientists did take it seriously, including the prominent psychologist Hans Eysenck (1916–1997), and this prompted some to undertake further tests of astrological effects, with uniformly negative results.[24]

Hans Eysenck: Another Sort of Nonsense

Scientists believe that fraud is rare in their profession, and they may believe it with good reason. It is rare not because of the extreme sanctions that the institutions of science apply but because it is difficult to prove fraud and dangerous to allege it without compelling evidence. If it is rare, it is because scientists are driven not only by external interests, not only by the gains they will accrue from career advancement, but because they are driven first by curiosity, by their "need to know." Deliberate falsehoods make a mockery of this need, a need that is at the heart of their perception of themselves.

Scientific fraud is deeply shocking to scientists, and they are unforgiving of it. It may be hard to prove fraud, but error, from whatever cause, within the small communities of scientists, is penalized by reputational damage. In the words that the novelist Dorothy Sayers placed in the mouth of her detective, Peter Wimsey, "The only ethical principle which has made science possible is that the truth shall be told all the time. If we do not penalize false statements made in error, we open up the way, don't you see, for false statements by intention. And of course a false statement of fact, made deliberately, is the most serious crime a scientist can commit."[25]

Eysenck had declared a commitment to scientific rigor, a commitment exemplified in his faith in statistical analyses that led him often to believe that correlations implied causality, and that abstract factors that featured in complex statistical analyses had concrete meaning. In this he followed the tradition of his PhD supervisor, Cyril Burt (1883–1971), best known for his studies on the heritability of intelligence, studies widely used to justify discrimination on grounds of race or class.

In 1974, with the publication of Leon Kamin's book *The Science and Politics of IQ*, Burt fell under suspicion of fabricating data to support his statistical correlations.[26] Other criticisms followed, of carelessness in his studies and of conceptual flaws in the interpretation of their outcomes. In 1995, Nicholas Mackintosh, professor of experimental psychology at the University of Cambridge, edited a collection of articles on the "Burt Affair."[27] In his summary of the evidence, he concluded that Burt's data were "so woefully inadequate and riddled with error [that] no reliance [could] be placed on the numbers he present[ed]."

In 2019, Eysenck's own studies fell under suspicion of fraud in his program of research into the causes, prevention, and treatment of fatal diseases.[28] In open letters to King's College London, where Eysenck had worked, and to the British Psychological Society, the editor of the *Journal of Health Psychology* requested a thorough investigation together with retraction or correction of sixty-one papers.[29]

Scientists all like to think of themselves as critical, skeptical, and rigorous, and perhaps they generally are, within their own particular field of expertise. But systematic skepticism is hard to sustain for a very human reason: we do not expect other people to lie or to deliberately mislead us. Language, that most human of attributes, would be valueless if we think that people lie lightly, so we trust what people say without good reason to do otherwise.

For the most part, people first deceive themselves, and then deceive others, by way more of collateral damage than intent. They make false statements in error.

11 Complexity and Its Problems for Causal Narratives

In the first half of the twentieth century, quantum physics and cosmology seemed to show, in their different ways, the impossibility of creating an understanding of fundamental aspects of the world that was compatible with our normal intuitions. In these realms, scientists might try to convey the spirit of their understanding to nonexperts, but not an understanding that could be applied in any meaningful way. This didn't seem to matter much—the paradoxes that these theories engaged lay outside the realm of normal experience. But other advances in physics posed a broader challenge.

Until well into the twentieth century, many of the basic theories of physics, such as Newton's theory of universal gravitation and Maxwell's electrodynamics, were essentially linear. In a linear system, "causes" have predictable "effects." For example, if a system is affected by two variables, then the sum of their separate effects is the same as the effect of the two variables applied together, if the system is linear. This is not true of nonlinear systems, so these are much harder to model.

Some nonlinear systems can be reasonably approximated by linear systems. However, many problems, such as those in aerodynamics and hydrodynamics, cannot be tackled this way.

Nevertheless, the second half of the twentieth century brought advances in understanding the ways in which nonlinear systems give rise to complex behaviors. For example, in a linear system, small changes in the values of variables have (relatively) small effects, but in a nonlinear system, changes that usually have small effects might, in some circumstances, have a large effect. A small change in the pressure applied to a beam or to a bubble might cause the beam to buckle or the bubble to burst. Because such phenomena are common in our normal experience, they seem amenable to causal narratives that do not unduly tax our intuitions. We must just be willing to recognize that factors that separately have little effect might, when present together, have a large effect.

Nonlinear systems behave in ways that, though complex and sometimes surprising, are nevertheless *deterministic*; from given "initial conditions" (the same starting values of all the variables) the system will behave the same way every time you test it. Because of this, it has until recently been assumed that understanding complex systems was essentially no different from understanding simple linear systems—they just needed more data, because they were, well, more complicated.

Advances in the mathematical understanding of nonlinear systems led to ways of classifying complex behaviors, with the promise that this might lead to inferences about the underlying causes, and scientists turned to using this to search for simple explanations of apparently complex phenomena. This search for simplicity was again driven by two things—by the need for utility in explanations and by the urge for them to have narrative potency. But in the 1970s came a shock with the recognition that fully deterministic systems might nevertheless be completely unpredictable in their behavior.

In 1963, a meteorologist, Edward Lorenz, published an article that has been described as "one of the great achievements of twentieth-century physics."[1] At first, few scientists other than meteorologists noticed it until, in 1972, he gave a talk to the American Association for the Advancement of Science. Lorenz was concerned with improving the reliability of long-range weather forecasts, and he had built a (relatively) simple mathematical model of weather systems. Implementing this on a computer involved setting "initial conditions" for the different variables, and his program simulated how a "weather system" described by these variables evolved over time from the initial state.

He later described his experience with this program:

> At one point, I decided to repeat some of the calculations in order to examine what was happening in greater detail. I stopped the computer, typed in a line of numbers that had come out of the printer a little earlier, and started it back up. I went to the lobby to have a cup of coffee and came back an hour later, during which the computer had simulated about two months of weather. The numbers coming out of the printer had nothing to do with the previous ones.[2]

He then recognized that the numbers that he had typed into the computer the second time were not exactly the same as the previous ones. They were "rounded" versions.

Lorenz's great insight was in understanding that these "errors" revealed a profoundly important truth. Virtually imperceptible differences in initial conditions can, over time, have a massive effect on the behavior of a complex system. This became known as the "butterfly effect," and the idea was propagated across another complex system—the internet.

Lorenz's 1972 talk was titled "Predictability: Does the Flap of a Butterfly's Wings in Brazil Set Off a Tornado in Texas?" Gareth once surfed the internet in search of an answer to this question.[3] A search for "Lorenz: butterfly effect" produced many hits to academic sites, and he looked at about fifty of them. Taking their assertions at face value, while butterflies in South America can cause typhoons in Asia, they might only "affect the weather" in Central Park. Brazilian butterflies can also cause hurricanes in Alaska, but here the trail died, perhaps because of a shortage of butterflies in Alaska. But Amazon butterflies can cause hurricanes in the Caribbean. If one should strike Cuba in August and disturb a butterfly there, then a flap of its wings might cause a hurricane in Florida in September, and Florida butterflies might, a week later, trigger a hurricane in Spain. However, if a Caribbean hurricane disturbed butterflies in Aruba, they might "completely change" the weather in Bali. Butterflies in nearby Tahiti could produce tornados in Kansas; what happens in Kansas may not matter much, but butterflies in San Francisco could "transform the weather conditions" in Honolulu or Shanghai, and Californian butterflies might ("eventually") "have an effect on the weather of the East Coast." New York butterflies could cause hurricanes in Japan, and Japanese butterflies hurricanes in America. Butterflies in "Cambridge or Boston" merely "affect the climate" in Europe two weeks later, which hardly seemed difficult if you count London as in Europe. However, London butterflies can affect the weather "on the other side of the world." Butterflies *anywhere* might trigger tornados in Indonesia, and butterflies in Java might cause the weather to "turn nasty" in Chicago. Chicago butterflies cause typhoons in China, triggering escalation as Chinese butterflies cause blizzards in Chicago, but Peking butterflies might "transform storm systems" in Nevada and "storm fronts could batter the shores of New York." New York butterflies might then cause hurricanes in Peking, and, if this escalation spreads, then, indeed, butterflies in Hong Kong might "change tornado patterns" in Texas.

Thus, aggregating the wisdom of the internet did support the claim that a butterfly in Brazil might cause a tornado in Texas. Many of these statements were attributed to Lorenz, and some were given as quotations, sometimes (always wrongly) attributed to his 1963 paper. Several sites quoted Lorenz as recalling that "one meteorologist remarked that if the theory were correct, one flap of a seagull's wings would be enough to alter the course of the weather forever." So perhaps the first butterfly was a seagull; but no, in 1898, W. S. Franklin, also writing about weather prediction, reportedly stated that its accuracy "is subject to the condition that the flight of a grasshopper in Montana may turn a storm aside from Philadelphia to New York!"[4]

Most sites recognized the butterfly story as rhetorical and that its purpose was to warn that nonlinear dynamical systems can be very sensitive to initial conditions. However, on most sites, the butterfly story (in one of its variant manifestations) was given as a statement, not as a question, and the prelude "Predictability:" was lost. Even where the title was given in full, the significance of the question mark was generally missed, because Lorenz's question was taken as evidently expecting the answer "Yes."

But Lorenz's question was intended as a question, and the word "Predictability" was also important. His question indicated that his model of weather systems posed a challenge to our understanding. If the model was any good at all, then weather patterns are inherently unpredictable. Accordingly, the butterfly statement is not a scientific statement because it is inherently untestable.

However, the surprise in the title carries a second important message. The statement "The Flap of a Butterfly's Wings in Brazil May Set off a Tornado in Texas" whether true or not, is nonsense, and it is nonsense because the notion of causality breaks down.

Classically, causality is central to scientific explanation: in the (reported) words of Thomas Hobbes, "Everything is best understood by its constitutive causes. For, as in a watch or some such small engine, the matter, figure and motion of the wheels cannot well be known except it be taken asunder and viewed in parts."[5] This assumption, that phenomena apparent at a higher level (systems, organisms) can be explained in terms of phenomena that operate at lower levels (cells, molecules), is commonly known as *reductionism*. But studies of cells and molecules have not always provided satisfactory explanations of phenomena at the level of the whole organism. There are several reasons for this. Complex systems may be so sensitive to initial conditions that everything we can know about the component elements will still not enable us to predict the behavior of the whole system. More tellingly, nonlinear systems that do not exhibit chaotic behavior can display organized and adaptive "emergent behavior" that only appears above a critical level of complexity, and which is accordingly a property only of the system as a whole and not of any subset of its component elements.

Here is the rub. Scientists, as Latour saw, are in the business of persuading people. Persuading journals to publish their papers, persuading the organizers of conferences to lend them a pulpit, persuading their peers to listen, persuading grant committees to fund them, and persuading the public that what they do is worthwhile. In all of these, the narrative potency of their explanations is paramount. We have to understand this. But we also must understand that narratives that are simple, clear, and memorable may speak more of our own cognitive constraints than of any underlying reality.

Or, as Richard Feynman put it when invited to explain his work on quantum physics for a general audience, "Hell, if I could explain it to the average person, it wouldn't have been worth the Nobel Prize."[6]

The insights from chaos theory have major implications for our concepts of predictability and causality. The simple narratives that scientists strive for do not convey, *cannot* convey, the nature of complex systems.

This is a challenge to the notion that scientific understanding approaches an absolute truth. The benefits of science come from its utility. A theory that is demonstrably untrue has no utility, but a theory that cannot be communicated effectively may have little utility even if true.

It also presents challenges for the scientific method. The conventional way of designing experiments assumes replicability. But how do we even go about investigating a complex system whose behavior is intrinsically unpredictable?

When A affects B, which affects C, which then affects A, talking of causes at all becomes arbitrary in the context of a complex system as a whole.

Science, by common accounts of its progress, seems like a linear system with a simple causal chain: paper A leads to paper B, which leads to paper C. But paper C can lead to a reinterpretation of paper A. Science itself is a complex system.

12   Publication and Citation: A Complex System

The frontiers of a book are never clear-cut: beyond the title, the first lines, the last full stop [. . .] it is caught up in a system of references to other books, other texts, other sentences: it is a node within a network [. . .] [it] is not simply the object that one holds in one's hands; and it cannot remain within the little parallelepiped that contains it.
—Michel Foucault[1]

Scientists propagate their ideas through the papers that they publish. It is only by engaging in this system, a complex system constructed and maintained by scientists for scientists, that they play the scientific game.

No scientific paper stands alone. Tracing the development of even an apparently simple scientific statement will uncover research that might span decades and involve hundreds of scientists from many disciplines. Collectively they might have published thousands of documents—primary studies, reviews, methodological papers, letters, opinion pieces, conference proceedings, books, abstracts, revisions, and retractions—and these are linked by a network of citations that is many times larger than the set of papers. This is a complex system of interactions that modify, reinterpret, qualify, embellish, and extend, previous work.

Communication between scientists, through their papers, enables a transition from the idiosyncratic to the communal. Through publication, scientists erect a vast edifice of data, hypotheses, and definitions, and, through citation, they acknowledge how they use this resource. But we understand only weakly how this system of communication influences the flow of ideas.

While few scientists would deny that Science relies on the work of those who came before, they are rarely fully aware of the impact of their own papers, nor are they good at predicting how a new paper will be received. Some papers rise to prominence soon after publication, others lie dormant for many years until, for some unforeseeable reason or accident, the scientific community takes notice, while others will be forever ignored. But most papers, make only a subtle impact.[2] Some may be read and have an impact without being cited; others are read and ignored; others are just ignored, and some are cited without having been read.[3] Why any particular paper becomes widely known and cited is sometimes obvious, but is often very hard to understand.

Scientific Publication

Science is a morass of different methodologies and theories. We classify very different kinds of research as scientific—from particle physics to anatomy—but there are few commonalities in the ways that different fields proceed. Few now take seriously the idea that sciences share a universal scientific method, except in a few particulars, but all sciences share the convention of publishing in peer-reviewed journals.[4] An idea or observation that goes unpublished is not part of the scientific knowledge base. Only through publication can the process of authentication and external critical scrutiny begin, and only then can a particular idea become widely known.

Scientific publication may be the only necessary factor for a piece of work to be considered scientific. As every PhD student knows, publication is often a prerequisite for gaining their doctorate, and the need to publish will continue for as long as they continue to play the scientific game. In 1942, Logan Wilson, in *The Academic Man*, introduced the term "publish or perish" to describe this pressure. The need to publish is not only about career advancement, it is also a moral imperative, for "in the academic scheme of things, results unpublished are little better than those never achieved."[5]

In 1945, Karl Popper reflected on what made science scientific, and while his thesis of falsification is well known, fewer noticed the importance he placed on communication. He asked his reader to imagine Robinson Crusoe abandoned on a desert island with all the equipment and resources necessary to conduct a scientific investigation, and asked: Can anything that Crusoe produces be called science? According to Popper, it cannot:

> For there is nobody but himself to check his results; nobody but himself to correct those prejudices which are the unavoidable consequence of his peculiar mental history; nobody to help him get rid of that strange blindness concerning inherent possibilities of our results which is a consequence of the fact that most of them research through comparatively irrelevant approaches. And concerning his scientific papers, it is only in attempts to explain his work to somebody who has not done it that he can acquire the discipline of clear and reasoned communication which too is part of the scientific method.[6]

Popper thought that publication was important not just to allow others to scrutinize the work, but also because the process of writing was in itself vital. This is perhaps true for all writing, not just scientific writing. In *Aspects of the Novel*, E. M. Forster recounted the anecdote of the old lady who, accused of being illogical, declared, "How can I tell what I think till I see what I say?"[7]

When scientists feel sure enough of their ideas and of the rigor of their experiments, they might write something—only to discard it, because it does not capture what they meant it to. When confident that what they have written has passed their own critical scrutiny, they may show drafts to colleagues and discuss them at departmental talks and conferences. They will amend their work in light of the comments and of the level of interest received. They will refine their manuscript to try to make it clear, memorable, and impactful.

They will omit details that are probably irrelevant. Omit experiments that didn't work. Filter the results to exclude "outliers" that might be misleading because some unknown confound corrupted them. Select experiments whose outcomes are most clear for illustration. Choose their methods of analysis carefully, to most cogently display a preferred interpretation. Unite their findings with a narrative that is memorable in being simple and direct, and discard findings that disrupt its flow. Tip their hats to the totems of the field; cite the established authorities. Cite those who they guess will be the chosen reviewers of the paper, and cite papers in high-impact journals to illustrate the importance of their work. Make bold claims for its potential utility.

The next hurdle is submission to a journal. For most journals, the paper will be considered, anonymously, by several independent referees. If they find errors, unsupported claims, or insufficient knowledge of other literature, it is likely to be rejected. It's an imperfect process: few referees have the expertise to judge forensically all the technical details of a paper. All have a partial, selective knowledge of the literature. They may look hard at papers that contradict common assumptions and less hard papers that "go with the flow." Their job is to provide constructive criticism, not to decide whether a paper will be accepted. It is the editor's task to weigh the often dissonant reports from referees and make a decision. If the decision is not "reject," it is most likely to be "revise": to amend certain passages, to fix problems with the reporting or analysis of data, to explain the methods more fully, or to extend the discussion in particular ways. Only after satisfactory revision will the decision be "accept."

For some journals, including most that aspire to have a high impact factor, there will be another hurdle, and whether this is crossed will be the judgement of a senior editor. If the paper is published, is it likely to add to the impact factor of the journal: will it be cited often?

Citation Behavior

How often a paper is cited is widely used as an indicator not just of influence but also of quality.[8] A good paper perhaps ought to receive more citations. Scientists ought to be able to judge the quality of papers in their field; as authors, they cannot cite all the relevant evidence, so they might be expected to cite the best evidence.

But a paper might be cited for many other reasons.[9] According to some critics, the objective "quality" of papers—their methodological rigor—plays only a minor role in how they are cited.[10]

Rather, authors use citations as rhetorical devices to persuade the community of the validity of their claims.

First, and most obviously, authors cite their own studies.[11] Much of this is natural and unavoidable. Few papers are self-contained; most build upon the authors' past work, the details of which would be tiresome to repeat. Authors also want to advertise their recent works, even if only tangentially relevant. It is not hard to find examples of excessive self-citation, and some journals now take a dim view of this. Men cite their own work more often than women do—according to a large survey, 56 percent more often.[12] It's hard to know what to make of this. Men publish more than women, so they might cite their own work more often just because there's more to cite. Alternatively, they might cite their work more often because they have read less widely, are more self-centered, or are more deviously manipulative in pursuing their ambition. Interpretation of any finding in the literature has considerable flexibility.

Second, authors cite studies that they are familiar with—those of their colleagues, friends, associates, and collaborators.[13] Again, it is not hard to defend this—these are likely to be papers whose strengths and weaknesses they are most aware of. However, this means there can be an entry bar to newcomers. A well-known author is more likely to have his or her paper accepted by a journal, and that paper is more likely to be cited by others in the field, than a paper from a newcomer who has still to persuade others of their competence.

Third, authors may cite the first study that reported a particular finding, even when later studies are much more complete. This follows from respecting priority in discovery.[14] This convention is considered important to uphold, if only to protect any claims to priority that the authors themselves might have. However, it is problematic, as we will see in chapter 14 in the case of citations of Henry Dale. One problem is one that the sociologist Barry Barnes called "meaning finitism": because "facts" are not fixed and immutable, because the meaning of terms changes in context and across time, the antecedents of a "fact" may have a contestable relationship to the current understanding of that "fact."[15] Another problem is that the earliest studies are unlikely to be the best. To establish priority, scientists tend to rush into print as soon as they have just enough evidence. Initial studies are often small, poorly controlled, and, for reasons to be explained later, likely to report inflated estimates of the effect size. There are greater problems in assigning precedence to ideas. Authors seek to acknowledge such intellectual debts, particularly where they are overt and contiguous, but as one of us commented elsewhere, "Science is a commonwealth, within which, by-and-large, knowledge, expertise, and ideas flow freely. Tracing the 'true' origin of any idea is like asking of a drop of water in a river from which cloud it fell, and about as useful."[16]

Fourth, authors cite recent papers—since the 1950s about half of all references are to papers published in the preceding five years.[17] Again, this is defensible; it is important to place any paper in the context of contemporary thinking. But it is also true that such citations are signposts to the editors of a journal. Authors cite papers written by those that they hope (or fear) will be asked as referees and expect to benefit from showing those referees that they respect *their* work. There is also evidence that some journals pressure authors to cite other articles published recently in that journal to inflate the impact factor.[18]

Fifth, authors cite papers that have been frequently cited.[19] Again this is defensible; authors seek to place their work in a context whose importance is well recognized. Highly cited papers attract more citations simply because they are highly cited.

Sixth, authors cite papers published in high-impact journals.[20] This comes from two incentives, from the wish to demonstrate that the topic of their own work is "hot," and from the perception that such papers are likely to be good papers because they have endured a rigorous selection process. Whether this is a sound perception or not is beside the point—to use a surrogate measure of quality in deciding which papers to cite is intellectually feeble.

No less important are the decisions that authors make of what not to cite. They tend not to cite studies that they consider weak or flawed. But they tend to cite studies with a positive outcome rather than those with a negative outcome, even if the latter are more rigorous. They tend not to cite inconvenient studies, with findings that would complicate or contradict their claims. They tend not to cite studies from little known authors or studies in less well-known journals. They tend not to cite studies from those with whom they have had strong disagreements (see Schally and Guillemin in chapter 5).

It is complicated—a scientist's reputation comes from peer recognition, and their peers will often recognize when a paper ignores conflicting evidence. Experienced authors handle these issues with delicacy—they cite their critics for things on which they are in agreement but not for matters in dispute. They use flexibility of interpretation creatively to interpret other studies as supportive wherever that can. They choose their words carefully in how they undermine the credibility of rival positions.

Nevertheless, the need to convince their peers helps to eliminate biases: "What we call 'scientific objectivity' is not a product of the individual scientist's impartiality, but a product of the social or public character of scientific method; and the individual scientist's impartiality is, so far as it exists, not the source but rather the result of this socially or institutionally organized objectivity of science."[21]

Publications and Citations: Impact

However imperfectly, a scientific paper pays homage to its antecedents and acknowledges current influences through the papers that it cites. Reflecting on his achievements, Isaac Newton wrote, "If I have seen further, it is by standing on the shoulders of giants."[22] By the popular interpretation, this was an assertion of humility, a reflection that any scientific advance relies on the work of many others. The convention of publication and citation captures this—it makes manifest the flow of ideas from one mind to another. In their papers, scientists claim priority for particular discoveries, and by their references they acknowledge their intellectual debts.

Citations, according to Eugene Garfield, are thus indicators of influence: "The total number of such expressions is about the most objective measure there is of the material's importance to current research."[23] In the 1950s, Garfield developed an approach, known as *scientometrics*, to develop this, and in 1960, he founded the Institute for Scientific Information (ISI) to index bibliographic data systematically. Today, ISI's Web of Science has fully indexed more than 73 million documents, including papers, conference proceedings, and abstracts, and their citations, a massive archive of publications and their interactions from the turn of the twentieth century to the present day.

While much of scientometrics is preoccupied with evaluating research performance, the origins of this field focused on the history and sociology of science and the desire to help scientists navigate the mass of information that confronts them.[24]

Certainly scientists cite works that have contributed to their own ideas, so the number of citations that a paper receives is, in part, a record of its influence. In addition, it is to be expected that studies conducted to a high standard will be cited more often than similar studies of a lower standard.[25] However, citations are also a tool of persuasion. For Latour, it was through the act of citing past literature that scientists constructed stories that could persuade their audiences:

> Whatever the tactics, the general strategy is easy to grasp: do whatever you need to the former literature to render it as helpful as possible for the claims you are going to make. The rules are simple enough: weaken your enemies, paralyse those you cannot weaken, help your allies if they are attacked, ensure safe communications with those who supply you with indisputable instruments, oblige your enemies to fight one another; if you are not sure of winning, be humble and understate.[26]

According to Latour, the scientific literature might be "cited without being read, that is perfunctorily; or to support a claim which is exactly the opposite of what its author intended; or for technical details so minute that they escaped their author's attention; or because of intentions attributed to the authors but not explicitly stated in the text."[27]

Even if this Machiavellian view is accepted, citations would still indicate the influence of particular papers. Indeed, the convention of citation appears to democratize science: authors choose what to cite and select those papers that they perceive as useful for advancing the claims that they are promoting.

Derek de Solla Price: Science as a Complex Dynamic System

Derek de Solla Price (1922–1983), a physicist who became interested in the history of science, saw in bibliometric data an opportunity to study Science in a new way. In 1963, he posed a simple question: "Why should we not turn the tools of science on science itself?"[28]

In the 1940s, the rebuilding of a library had left Price in custody of a set of the *Philosophical Transactions of the Royal Society of London* from 1665 to the 1930s. He sorted these into piles by decade and noted something odd: the piles conformed to an exponential growth curve. The difference between linear and exponential growth is the difference between simple and compound interest. An investment of $100 that gains a simple interest return of $10 per year will grow linearly to $300 in twenty years. But that same investment with a compound interest return of 10 percent per year will grow exponentially to more than $600 in twenty years. A feature of exponential growth is that is has a constant "doubling time": the investment of $100 doubles about every eight years, to $200, $400, $800, and so on.

Price followed this by examining physics abstracts published between 1900 and 1950. Disregarding the dips in productivity during the world wars, their number had also been growing exponentially, with a doubling time of about ten years. Abstracts published on the theory of determinants and matrices grew similarly, doubling about every twelve years between 1760 and 1950. Price recognized that exponential growth was a property of systems in which the rate of increase at any time is proportional to the amount already produced: "In the two cases studied, this constant of proportionality is such that the whole of the previous work could have been done in about sixteen years if publication had proceeded constantly at the current rate."[29]

But, according to Price, his paper went down like a "lead balloon." Undaunted, he followed up by studying the entire scientific enterprise and found exponential growth in the number of papers, the number of scientific journals, and the number of professional scientists, which seemed also have a doubling time of 15–20 years, Scientists, like the papers they produce, appeared to be multiplying steeply and with remarkable regularity.

By this estimate, about 80–90 percent of all scientists that have ever lived are alive now, and this has been true since the mid-sixteenth century. This meant that Science as experienced by scientists always feels modern—a view that appeared at odds with the main focus of historians of Science, whose attention was on "giants" such as Newton, Maxwell, and Darwin. Indeed, it seemed that, for most scientists, most of the work important to them had been published within about the last five years.

Price's findings have been replicated many times. Lutz Bornmann and Rüdiger Mutz looked at the growth of science since 1650 and noted that, since the end of World War II, the size of the scientific literature has been doubling every 9 years.[30] This is not because individual scientists have been more productive—their rate of publication has remained unchanged for over a century. The growth in scientific output matches the growth in the number of scientists.[31]

This feature of exponential growth in the volume of science doesn't fit easily with our intuitive conceptions of science as a linear progression of our understanding. Rather, it seems to show science as constantly expanding, in all available directions and in all available ways, as the number of scientists increases.

Citation Networks

Price reflected on what implications this might have for scientific knowledge and turned his attention to the convention of citation.[32] For a large sample of papers published in 1961, he surveyed the list of papers that they had cited. A few papers had been cited by many of the 1961 papers. Overall, the distribution of citations was very uneven: it conformed to a *power-law distribution*— the proportion $P(n)$ of papers with n citations is proportional to n^{-p} for some fixed power p. Price saw that "in any given year, about 35% of all existing papers are not cited at all, and another 49% are cited only once." This doesn't mean that they weren't read, but their influence, if any, was invisible. Just 1 percent of papers were cited six times or more.

The world of science in 1961 was different from that of today. Reference lists were smaller, and Price's publication lists included abstracts that were rarely cited and had few references. In chapter 13, we look at one field by surveying just primary research papers. In every decade we analyzed, these conformed to a power-law distribution as Price had described—and to a virtually identical distribution in each decade. Today, while exactly what kind of model best fits citation data is a topic of dispute, heavily skewed distributions of citations have been found in every field thus far examined.[33]

This striking and apparently universal feature of citation networks was unexpected, and seemed to suggest that the flow of knowledge through these networks was very uneven. There seemed to be a strong, uniform selection pressure on the literature that determined how, in all fields and at all times, just a few papers would be disproportionately influential. It thus seemed important to understand how the power-law distribution arises.

Cumulative Advantage: The Matthew Effect

In 1976, Price sought to understand the rules that governed the inequality in citations:

> Success seems to breed success. A paper which has been cited many times is more likely to be cited again than one which has been little

cited. An author of many papers is more likely to publish again than one who has been less prolific. A journal which has been frequently consulted for some purpose is more likely to be turned to again than one of previously infrequent use. Words become common or remain rare. A millionaire gets extra income faster and easier than a beggar.[34]

This attracted the attention of Merton, who proposed that scientific success and failure can be summarized from a passage in the Book of Matthew: "For unto every one that hath shall be given, and he shall have abundance: but from him that hath not shall be taken away even that which he hath."[35] For Merton, the citation system was a system of peer recognition, whereby those already well known became ever better known—"a scientific contribution will have greater visibility in the community of scientists when it is introduced by a scientist of high rank than when it is introduced by one who has not yet made his mark."[36] Two explanations seemed possible: (1) scientists respect well-known scientists, preferentially citing their work; and (2) the Matthew Effect increases the visibility of contributions from well-known scientists and reduces that from less well-known scientists. The first explanation fits with the "normative" account of citation—that references reflect perceived quality; the second suggests that mere "visibility" is important.

Price tried to model this distribution:

> the model supposes that fate has in storage an urn containing red and black balls; at regular intervals a ball is drawn at random, a red ball signifying a "success" and a black ball a "failure." If the composition of the urn remained fixed the chances of success and failure would not vary, but if at each drawing the composition is changed by some rule, the chances will change as an after-effect of the previous history.[37]

The Matthew Effect suggests that success will attract success, so:

after each drawing the ball is replaced; if a red is drawn then c red bails are added, but if a black is drawn no extra balls are put in the urn. If we start with b black balls and r red, the conditional probability of success after n previous successes will be $(r + nc)/(b + r + nc)$ and the corresponding conditional probability of failure will be $b/(b + r + nc)$.[38]

By this account, new papers preferentially cite papers that are already well cited. This simple model fit empirical data on actual citation distributions, so Price concluded that citation dynamics could be explained by two simple functions—cumulative advantage and population growth.

In Price's time, few scientists were aware of the number of citations made to any particular paper: the ability to quickly see how many citations an article has received is a recent development. Merton's idea that citations increase the visibility of papers seems to be sharp—highly-cited papers are noticed more often because they are mentioned in more papers, and as a result they are cited even more often.

If the number of publications and citations produced depends mainly on the number of scientists working on that subject, if only a few papers are highly cited, what are the implications for our understanding of Science? In the following chapters, we explore some of them.

13   A Case Study of a Field in Evolution: Oxytocin, from Birth to Behavior

> Every discipline, in fact almost every problem, has its own vanguard, the group of research scientists working practically on a given problem. This is followed by the main body, the official community. [. . .] The vanguard does not occupy a fixed position. It changes its quarters from day to day and even from hour to hour. The main body advances more slowly, changing its stand—often spasmodically—only after years or even decades. [. . .] This indubitable phenomenon is obviously social in character and has important theoretical consequences. If a scientist is asked about the status of a given problem, he must first specify the [textbook] view as something impersonal and comparatively fixed, although he knows full well that it is inevitably already out of date.
>
> —Ludwik Fleck[1]

To understand how a field of science develops, how ideas spread and change, we might try to look at all the papers in a field. These, by the papers that they cite, acknowledge their debts to previous papers, and their reach can be seen in the papers that cite them. Citation networks trap the paths by which facts are created.[2]

But the output of science is vast. The Web of Science indexes papers published in more than 20,900 journals, books, and conference proceedings. It contains records of about 73 million articles and of more than a billion citations to them.

Each field of science has its own pattern of citation, a pattern that changes over time. Once, all journals were published in print, and because printing was expensive, journals restricted the number of references that any paper could cite. With online publication, authors no longer need to be so selective, and as a consequence, recent papers appear to be more impactful in being cited more often. That there are also differences between fields is unsurprising: papers in mathematics tend to have few references—they are largely self-contained. At the other extreme, papers in biology often refer to a hundred or more papers as authors relate their work to a rapidly growing body of findings.

If you ask any scientist what the important papers are, they might reflect on what papers have influenced them, or on which have had the biggest societal or economic impact, or the biggest methodological impact. If you ask which are the best papers however, the answer will probably be different, and different for each scientist.

If we ask which papers have been most cited, the answer will be objective. How we interpret that answer is something else, but we might expect the most cited papers to include many that most scientists in the field would regard as important and many that most would regard as the best. What we cannot do is to judge one field of science against another in this way, nor can we judge papers of one time against those of another.

The more a paper is cited, the more it is noticed and the more citations it gets. Looking at the most highly cited papers at a given time gives some idea of what ideas were considered most important in subsequent years.

Our purpose here is to look at a single field, one with a long history, to understand in what sense there has been progress and how it has been achieved. We looked at papers with "oxytocin" in their titles. The field is relatively small, but oxytocin is well known to many, especially as something commonly given to women in childbirth. Not all papers in the field mention oxytocin in the title, but many of those for which it is particularly salient do.

By May 2019, Web of Science had indexed more than 14,000 such items. These include 9,211 papers in peer-reviewed academic journals. By May 2019, these had been cited more than thirty times each, on average. About a third of all these citations come from other papers with "oxytocin" in the title: this search has not captured the whole oxytocin literature, but a large part of it.[3]

The field has grown unevenly (figure 13.1). From the 1950s to the early 1980s there was steady growth, followed by a "bulge" in productivity with a peak in about 1992, falling to a nadir in about 2005, then a further escalation. So, what is behind this pattern?

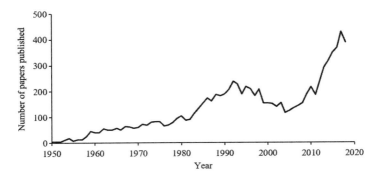

Figure 13.1
The number of papers with oxytocin in the title published each year, including primary research papers and review papers.

The Birth of the Birth Hormone

The discovery of oxytocin is often attributed to a British physiologist, Henry Dale. In 1906, he published a study of ergot, a herbal remedy once used by midwives to help deliver the afterbirth, and mentioned, as an aside, that an extract of pituitary glands could stimulate uterine contractions.[4] In 1909, William Blair-Bell tested it on pregnant women and found dramatic effects on labor.[5] Within another four years, it was in widespread use in obstetric clinics throughout North America and Europe.[6]

In 1928, Oliver Kamm, a chemist at Parke-Davis & Co., produced two, more refined extracts: one increased blood pressure and reduced urine production; the other promoted uterine contractions and stimulated milk letdown.[7] He hypothesized that each was enriched in a different "principle." He named the hypothesized principles "vasopressin" and "oxytocin" (from the Greek for "quick birth"), and the two extracts, *Pitressin* and *Pitocin*, became available for clinical use.

Between 1928 and 1950, fifty papers mentioned oxytocin or Pitocin in the title. Most had tested Kamm's postulate, either by exploring the effects of Pitocin on the uterus, or by comparing its cardiac effects with those of Pitressin. By 1950, oxytocin was accepted to be a *hormone*—a chemical messenger produced at one site in the body that is carried, by the blood, to act on tissues elsewhere in the body.

Oxytocin the Chemical

Between 1950 and 1959 oxytocin was in the title of ninety-four papers. Twenty-eight of these have received at least forty citations, including nineteen chemistry papers.

The extracts were still crude. In 1955, Derek Llewellyn-Jones wrote "there is no place for its use as a routine in the initiation or stimulation of labour; and there is little place for its use alone in the induction of labour."[8] His paper was cited three times in the following year and never again; it was made obsolete because chemists had identified the amino acid sequence of oxytocin and used this to synthesize pure oxytocin. For these advances, Vincent du Vigneaud was awarded the 1955 Nobel Prize in Chemistry, and the most highly cited paper of the decade (cited 944 times) is one of his.[9] Pure oxytocin became part of the routine management of labor.

In this decade there were many other advances.[10] For example, it became accepted that oxytocin was not made in the pituitary, but by neurons in the hypothalamus and that it was released by electrical signals in these neurons. "Oxytocin" was not in the titles of these papers; for many researchers, oxytocin was still a mystery—it had effects, but their nature and origin were poorly understood. But after du Vigneaud, oxytocin was a "thing," something that could be made in the lab—something that could, in principle, be measured directly by what it was, not indirectly by what it did. Something that could, in principle, be "seen." With this increased salience, "oxytocin" became a prominent inclusion in the titles.

Oxytocin the Drug

Between 1960 and 1969, 325 papers had oxytocin in their titles. The top sixty-two have each been cited more than forty times, and chemists produced seventeen of them. The search was now on to develop drugs more potent than oxytocin itself. The most highly cited paper described the synthesis of one potent variant of oxytocin.[11]

But pharmacology was taking over from chemistry: pharmacologists are concerned with the utility of drugs—how they might be used therapeutically. Twenty-four top papers were studies of oxytocin's effects on the uterus, on the kidney and bladder, on the ovarian cycle, on cardiovascular effects, and on metabolic effects. One paper reported effects on sexual behavior in male rabbits.

Chemists and pharmacologists published in different journals, but there was a symbiotic relationship. Pharmacologists needed chemists to supply them with drugs, and chemists needed pharmacologists to understand the biological properties of those drugs. Drugs were not generated at random, nor in response to a therapeutic need—chemists synthesized drugs that could be characterized by pharmacologists. Understanding the effects of a drug required a clear response that could be measured in a standard way, such as a change in blood pressure or a change in urine flow or contractions of a uterus. The direction of progress was determined by what could be done, not what "needed" to be done.

Indeed, it was not clear what, if anything, needed to be done. There is no disease known that is linked to either a deficiency or an excess of oxytocin. Oxytocin could help in labor and perhaps in breast-feeding, but it was hard to see how the pure oxytocin could be improved on for this. Pharmaceutical companies had little interest, except in the possibility that oxytocin antagonists might be used to prevent premature birth. The chemists and pharmacologists that studied oxytocin were driven by no commercial interests, nor by any clear goal. For the chemists, there was an abstract challenge—what could the structure of a biological molecule tell us about its biological function? For the pharmacologists, this was a hunt for some potential utility, but also for clues to an understanding of what oxytocin was actually used for in the body.

Oxytocin the "Neuropeptide"

Between 1970 and 1979, 521 papers had "oxytocin" in their titles. The ability to synthesize oxytocin had led to the ability to produce antibodies to it, and this gave rise to new technologies that enabled studies to move beyond pharmacology to physiology.

First, it became possible to measure oxytocin using *radioimmunoassays*—a technique for which Rosalind Yalow shared a Nobel Prize in 1977. Many of the top papers used this technique; the most highly cited of them reported evidence of oxytocin release in the brain.[12] Radioimmunoassays were no more sensitive than the bioassays that they were beginning to replace, but they could measure oxytocin in many samples at a time. Many laboratories specialized in assays and maintained them for a community of collaborators—few laboratories had the resources to both run a high-quality assay and generate enough samples to make the assay worthwhile.

Second, it became possible to "see" oxytocin in cells using *immunocytochemistry* to visualize exactly where antibodies to oxytocin bind, and the most cited paper of the decade (with 756 citations) reported that, while most oxytocin neurons send nerve fibers to the pituitary, some send fibers to other sites in the brain.[13] Oxytocin, it was inferred, is not just a hormone but also a neuropeptide, and many began to describe it as a neurotransmitter.

Another eight top papers reported the first electrophysiological studies of oxytocin neurons. The most highly cited of these revealed that, in response to suckling, the neurons discharge in bursts of electrical activity resulting in pulses of oxytocin secretion.[14] The introduction of electrophysiology changed our understanding of what had been thought to be a simple physiological reflex.

With these developments, neuroscientists began to dominate the oxytocin field.

Oxytocin the Love Hormone

Between 1980 and 1989, 959 papers were published, and 135 of them have been cited more than a hundred times. The most cited paper reported that oxytocin can stimulate maternal behavior.[15] A pregnant rat will pay little attention to any pups that are placed in her cage, but after giving birth, she will build a nest into which she will gather any pups placed in her cage. This behavior can be triggered in virgin rats by injecting oxytocin into the brain. This surprising finding was challenged: two papers reported their failure to reproduce the results, but others confirmed and extended them. Others reported that, in sheep, oxytocin released into the brain during parturition triggered the formation of the bond between a ewe and her lamb—maternal "love."[16]

Other top papers report that oxytocin is secreted in response to stress and food intake, as well as during masturbation; it facilitates sexual behavior in female rats, induces penile erection in males, suppresses appetite, reduces pain, increases gut motility, and regulates prolactin secretion.

The development of oxytocin agonists had enabled oxytocin receptors to be localized and their abundance measured. Receptors were found in some places where no oxytocin was present, and this raised questions about whether oxytocin is indeed a neurotransmitter—a *neurotransmitter* is, by the classical definition, something released by one neuron that acts on a directly adjacent neuron. A *hormone*, on the other hand, is released into the blood and acts on distant targets. Neither definition meets the circumstance that oxytocin may be released at particular sites in the brain but act at other, distant sites within the brain. Scientific definitions, even of such fundamental elements as a hormone or a neurotransmitter serve a purpose in a particular time and a particular context, but times and contexts change.

Little of this research appears to conform to "normal science" as described by Kuhn, or to the hypothesis testing of Popper. Most was engaged in discovering new roles for oxytocin. There were many theories, but little sign of a determined assault on them, except in the case of maternal behavior, where the outcome was inconclusive. Hypothesis testing was apparent in the design of many experiments, but scientists seemed to be arguing more with themselves rather than against any external actors; the reader of these papers is cast into the role of an observer of an ongoing debate within the minds of the writers. Scientists were probing—exploring possible avenues for future research and trying to build a case for them with papers that sought to answer anticipated objections.

Forty-four top papers are concerned with effects of oxytocin on the uterus. A new focus of attention was the oxytocin receptor. Oxytocin is secreted during parturition, but its effects depend on the receptors in the uterus, which increase massively in number towards the end of pregnancy. It was easy to measure oxytocin in rats during parturition, but circulating concentrations are much lower in women than in rats. However, in women, this is compensated for by the extreme sensitivity of oxytocin receptors at the end of pregnancy—the increase in sensitivity is so great that it is still unclear whether any increase in secretion is needed for oxytocin to trigger birth in women.

The "Bulge"

To see how the composition of the field has changed, we can use the categories by which the Web of Science classifies papers. The "bulge" between 1982 and 1992 is largely accounted for by papers classified as neuroscience or as endocrinology and metabolism. By contrast, papers in obstetrics and gynecology appear at a steady rate (figure 13.2).

In the late 1970s, two new techniques were introduced—*radioimmunoassays* and *immunocytochemistry*. The first made it easier to measure hormones in blood samples, the second made it possible to visualize cells that expressed peptides.

If we look at similar fields over this period, we can see a similar bulge (figure 13.3). Prolactin and luteinizing hormone are

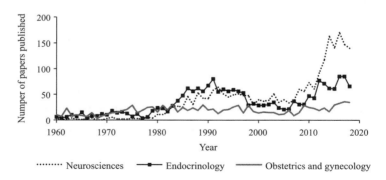

Figure 13.2
Numbers of "oxytocin" papers published each year, classified as neuroscience, endocrinology and obstetrics & gynecology.

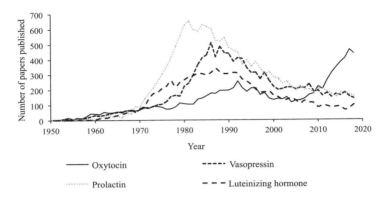

Figure 13.3

The number of papers published each year with "oxytocin," "vasopressin," "prolactin," or "luteinizing hormone" in the title

larger peptides than oxytocin, easier to raise antibodies to, and they are at higher concentrations in the blood, so assays for these become used more intensively. Vasopressin is very similar to oxytocin, but was considered more important because it is linked to many diseases. But for these and other pituitary hormones, the profile is similar—a bulge in publications associated with the use of radioimmunoassays. Not all of this rise is due to papers that use radioimmunoassays, but the findings with radioimmunoassays raised questions that stimulated other work.

In all cases the bulge subsided by the early 2000s. Radioimmunoassays are becoming old hat—the easy returns have been realized. The appearance of a decline in the field is misleading; in the 1980s there was a burst of exploratory activity, with new roles of oxytocin in diverse areas. These are now being pursued in other fields—the appetite field, the stress field, and others—and "oxytocin" seldom appears in the titles of these papers.

Oxytocin the "Cuddle Hormone"

Between 1990 and 1999, 1,357 "oxytocin" papers were published, and ninety-one have been cited more than a hundred times. The most cited paper reported the structure of the human oxytocin receptor, but at least fifty-five top papers were concerned with the actions of oxytocin in the brain, and seventeen of these addressed its role in social, maternal, or sexual behaviors.

That oxytocin could promote maternal bonding had prompted the question of whether it was also involved in partner bonding,[17] and attention turned to one of the few mammalian species that is socially monogamous. When a male prairie vole meets a receptive female, they engage in repeated coitus over about two days. This forges a long-lasting bond that is easy to recognize—pairs of prairie voles typically spend their time side by side; the voles huddle together rather than cuddling, but cuddling is what the press called it. When the young are born, the male plays his part in their nurture, and although voles vary in sexual fidelity, the social bond between partners is lifelong.

Three papers reported that oxytocin, given into the brain, can facilitate the formation of partner bonds. The *partner-preference test* was simple. A female vole was introduced into a cage where she had access to two other compartments, each of which housed a male. Would she spend her time with the male she had mated with a week before, or with a new male? Mating would produce a bond—but so would oxytocin.

These protocols could be readily replicated, and could be combined with mechanistic studies. Promiscuous and monogamous species had oxytocin receptors in different brain regions. The bonding was different for males—they became territorial and aggressive towards other males, and this involves vasopressin, not oxytocin. The brain regions involved were identified, and the vasopressin and oxytocin receptors were mapped.

These studies attracted enormous media attention. They also engaged the attention of the most media-savvy of scientific communities—psychologists.

Oxytocin, the Trust Hormone

Between 2000 and 2009, 836 "oxytocin" papers were published, including "Oxytocin increases trust in humans."[18] This, the most highly cited oxytocin paper to date, reported effects of intranasal application of oxytocin on how people play a "trust game" with monetary stakes where success requires the player to develop trust in a playing partner.

There was nothing new about the trust game; it was an established part of the repertoire of behavioral economics—the study of how people make economic decisions. Nor was there anything new about intranasal oxytocin: in the 1960s it was widely used to facilitate childbirth but had fallen into disuse as intravenous drips were more reliable.[19] This paper gave life to the notion that by a simple intervention—an intranasal spray—it might be possible to treat social behavior disorders. It used a very high dose— 24 international units (IUs), close to the total pituitary content of oxytocin in men, and more than twice as much as is needed to facilitate childbirth. This became the standard dose for several hundred studies in humans, examining psychological effects and effects on brain function.

In this decade, eighty-nine papers had, by May 2018, been cited at least a hundred times. Of these, twenty-five are studies of intranasal oxytocin in humans. Six are studies of genetic variation in the oxytocin receptor, identifying associations with autism, parenting behavior, and stress reactivity. Five are studies of plasma oxytocin concentrations in humans, linked to autism, sexual dysfunction, depression, maternal behavior, and social contact.

Forty-nine top papers are studies in animals, and of these, just one used intranasal oxytocin. Nevertheless, thirty-six papers are about behavioral actions of oxytocin: seven about maternal behavior and four about pair bonding. Seven papers are about social recognition, including a report that transgenic mice that lack oxytocin receptors cannot distinguish between familiar and unfamiliar mice. Sixteen are about stress, and eight of these on how oxytocin reduces anxiety.

Six top papers are on the metabolic roles of oxytocin—five on appetite, and one on bone growth. One study reported effects of oxytocin on heart muscle, and one that oxytocin and its receptor are synthesized in blood vessels. Just three top papers are studies of human labor.

Oxytocin the Intranasal Hormone

Between 2010 and May 2018, 752 "oxytocin" papers were published and had already amassed 24,215 citations. Of these, 143 were studies of intranasal oxytocin in humans and another eleven of intranasal oxytocin in animals.

In these years, eighty-five papers concerned effects on appetite and energy expenditure, including studies on bone, fat tissue, and on the gut. Twelve addressed actions of oxytocin on the heart. Eleven papers addressed its role in pain—the neurons in the spinal cord that carry pain messages are innervated by oxytocin neurons. But another 262 studies addressed the actions of oxytocin on the brain, including its effects on anxiety, reward processes, fear, aggression, maternal behavior, partner bonding, sexual behavior, and penile erection.

The effects of oxytocin on social behavior have triggered the interest of behavioral scientists, psychologists, and psychiatrists—few "oxytocin" papers came from these communities before 2009 (figure 13.4). As seen in figure 13.2 there was a corresponding rise in neuroscience papers. The interest in human behavior is driving (and being driven by) increasing attention to the neural mechanisms by which oxytocin influences behavior. The two communities rarely interact directly, but they are symbiotic.

The measurement of oxytocin has also returned, not in the context of endocrinology but in the context of psychology. These newcomers are using commercial assay kits that are unsuited to measuring plasma oxytocin, and understanding of the problems has been lost.

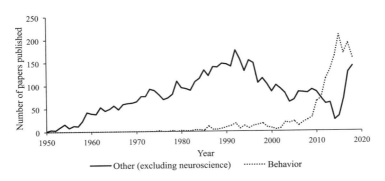

Figure 13.4
The number of papers published each year with "oxytocin" in the title in two classes: "Behavior," which combines papers in psychology, psychiatry and behavioral sciences, and papers in all other areas of study(-excluding neuroscience).

As also seen in figure 13.4, a new community, clinical neurology, is looking at the effects of intranasal oxytocin on brain activity. This community assumes that if there is a change in brain activity after intranasal oxytocin, then the oxytocin must have reached the brain. Disconnected from the researchers that had studied peripheral actions of oxytocin, it seems unaware of their existence.

The data thus shows that in the 1960s the oxytocin field was dominated by chemistry and pharmacology, in the 1970s by physiology, in the 1980s by endocrinology, in the 1990s by cellular neuroscience, in the 2000s by behavioral neuroscience, and in the 2010 onwards by psychology and psychiatry. Yet, as seen in figure 13.5, the field seems remarkably stable in the pattern of citation behavior. This graph shows how often research papers with "oxytocin" in the title have been cited, cataloging

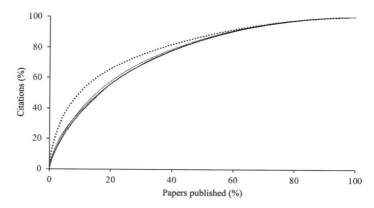

Figure 13.5
How often primary research papers with "oxytocin" in the title have been cited, for papers published in each of the five decades from 1960 to 2009. The search was conducted in May 2018.

papers published in each decade from 1960 to 2009.[20] In each decade, about 30 percent of the papers received about 70 percent of all citations—very like the distribution seen in other fields.[21] The lines for four of the decades are so similar that here they are indistinguishable as separate lines. The exception is the dotted line—papers published between 2000 and 2009, a decade in which 829 papers were published that had received 39,894 citations by May 2018. In this decade, which included the most highly cited paper ever in the field, the top 30 percent of papers attracted 75 percent of all citations.

The field has what mathematicians call a "fractal structure." Whether you look closely at one subfield, for example, electrophysiology, or stand back to the neuroscience of oxytocin, or further back, at the whole oxytocin field, or further back still, to the whole of neuroscience, the citation characteristics follow much the same pattern. That pattern is the same if you start with a different small subfield or if you start with a theme.

This pattern arises from the self-organizing structure of science. No scientist is an island, they work in communication with those who use a similar methodology and have similar interests. They cite mainly their nearest neighbors in the complex topological space of their science, and these small clusters provide the methodological and conceptual checks and balances. They cite others less directly, often either through reviews or via the "totems" and authorities of fields—the hubs by which ideas and innovations spread across communities. Remarkably, their citation behavior produces a virtually identical distribution of citations in every decade.

What appears striking is the absence of "goal-directed activity." Rather, scientists are opportunistic. New drugs, methodological advances such as radioimmunoassays and immunocytochemistry, and new experimental models such as intranasal application of peptides, open up possibilities not for solving existing problems but for new lines of inquiry.

How "Big" Is the Oxytocin Field?

The 1,357 research papers published between 1990 and 1999 had 2,852 authors. Of these, fifty-seven had each authored at least ten papers, and together they accounted for 584 papers—43 percent of the total. These fifty-seven authors were not independent— more than 200 of the 584 papers were collaborations between two or more of them. They also authored many review articles and were invited to present their work at specialist conferences. They were among the main "opinion leaders" in the field, but were distributed amongst several specialties, including obstetrics, neuroscience, endocrinology, pharmacology, and behavior. There are many different "areas" of the field, but each has just a few leaders whose influence is disproportionately great. Their ideas and insights spread, but so do their prejudices and misconceptions.

As an area grows, its citation rates increase. This is because most citations are made to closely related papers: in a new, rapidly growing area, the number of papers published at any time will be large compared to the number of "citable" papers, so the average citation rates will be high. This attracts recruits and grant income, as it is taken as a sign of quality and importance, and leads to further rapid escalation of activity. Eventually the quick returns diminish and it becomes harder for newcomers to make their presence felt; established fields have lower citation rates as the body of citable literature is large and grows slowly in proportion to its size.

Reflections

Oxytocin is only *essential* for milk ejection, but the understanding of what it is *important* for has changed. Once the hormone of "quick birth," oxytocin became the "love hormone" of sexual activity and maternal behavior, the "cuddle hormone" of partner bonding, then the "trust hormone" of social behavior.

In the account we can recognize technological milestones. The isolation of purified extracts and their application to establish some of the peripheral effects of oxytocin. The identification of the structure of oxytocin. The development of radioimmunoassays and immunocytochemistry. The sequencing of the oxytocin gene and of its receptor.

We can also recognize many discoveries. For example, oxytocin is produced by neurons, not by endocrine cells; its secretion is determined by electrical activity; and oxytocin neurons communicate to other neurons in the brain. In a sense, these are now accepted facts, but none is immutable or unqualified. Oxytocin is not only produced by neurons; its secretion from neurons is not only determined by electrical activity; and all oxytocin neurons do not communicate to other neurons in the brain.

As the oxytocin field developed, the theories that guided it mutated. The recognition that oxytocin receptor expression is dynamically regulated turned Harris's ideas (chapter 5) on their head. He had put the brain at the top of the regulatory hierarchy of the endocrine system, but if the actions of hormones are also governed by receptor expression in the target tissues then the site of control becomes unclear. In parturition, the uterus controls the brain as much as the converse.

But there seemed to be no overarching theoretical edifice that dictated what the important questions were and no prescriptive edifice that determined how questions may be answered and how the apparent answers should be interpreted. Rather, important questions were determined by new experimental tools. This was driven by the need of newcomers to a field to invest in expertise that would be in continuing demand and to be part of a community small enough to welcome them, but large enough to be likely to cite their papers extensively. This is not to say that scientists don't care about possible utility, but this is just one of many factors that influences the popularity of a field. Nor is it to say that scientists are not inspired by grand visions: many are, at least when grand visions are articulated by scientists whose work has been cited often.

We might see, in the diversity of oxytocin's effects, in the small communities that struggle to understand each feature, to tell the important from the trivial, to grapple with incongruities and inconsistencies only visible to them, that this is not a field at all but an ecologically diverse landscape.

14 Where Are the Facts?

Thoughts pass from one individual to another, each time a little transformed, for each individual can attach to them somewhat different associations. Strictly speaking, the receiver never understands the thought exactly in the way that the transmitter intended it to be understood. After a series of such encounters, practically nothing is left of the original content. Whose thought is it that continues to circulate? It is one that obviously belongs not to any single individual but to the collective. Whether an individual construes it as truth or error, understands it correctly or not, a set of findings meanders through the community, becoming polished, transformed, enforced, or attenuated, while influencing other findings, concept formation, opinions, and habits of thought.

—Ludwik Fleck[1]

There are few, if any, objective scientific facts, in the sense that those who want to "just know the facts" would be satisfied with. There are "facts" in the literature in the sense of the many uncontroversial and uncontested assertions about what has been observed in the course of experiments and about how those experiments were conducted and the observations derived. But if we look for what such "facts" mean, we will find only statements, the interpretation and valuation of which are disputable.

Scientists seldom talk of "facts," but of findings, observations, and conclusions. When they speak of facts, it is generally a rhetorical device, often an expression of impatience, to distinguish between statements that have an extensive foundation in empirical observations and statements that are speculative or disputed. Yet even such "facts" are always provisional, always subject to refinement, qualification, and, ultimately, rejection. This does not mean that science does not progress, only that its progress does not involve an accumulation of absolute facts.

The evolution of science is a process of refining, qualifying, modifying, reconfiguring, and replacing "facts"; this involves as much an expansion of what we know we don't know as an expansion of what we think we do. That this is indeed progress might be inferred from our success in applying what we know to practical ends, such as more effective ways of managing childbirth. For scientists, however, progress is generally less tangible, more understood as the evolution of a fuller mechanistic understanding that has greater predictive power in the experiments that they are then able to devise. That evolution involves a continual reinterpretation of data, reassessment of their validity, and reformulation of basic concepts and theories.

The only real facts in any story in science are statements about what has been said in the scientific literature. These we might reasonably call "facts" because anyone can confirm that what is stated to have been said has indeed been said by going to a library. If we see the same statement repeated often in apparently authoritative articles, we are perhaps unlikely to question it. But we should.

Who Discovered Oxytocin?

The *Encyclopedia Britannica* and many other sources declare as a "fact" that "the existence of a neurohormone with effects on uterine muscle tissue was demonstrated in 1906, when English physiologist Sir Henry Dale found that extracts of posterior pituitary glands from oxen, when administered to animals such as cats and dogs, encouraged the uterus to contract."[2] Wikipedia expresses it more bluntly: "Oxytocin was discovered by Henry Dale in 1906."[3]

It is true that, in 1906, Dale published a study of how ergot affects the uterus, and in that paper he mentioned that an extract of ox pituitary could also stimulate uterine contractions in cats.[4] He used an extract of the whole pituitary, not one of the posterior pituitary. What led Dale to test it on uterine contractions is not clear; he gave no explanation and no methodological details and mentioned the finding only in passing. It seems possible that he mentioned it only to establish priority, as one of his contemporaries, William Blair-Bell, had expressed interest in testing the effects of the extract.

However, Dale did not propose the existence of a "neurohormone." At that time, there was no suggestion that anything secreted from the pituitary had a neural origin. Even the concept of a "hormone" was new, and he did not use it. Dale believed that the actions of pituitary extracts on the uterus were those of a "pressor" principle that raised blood pressure (corresponding to the hormone later identified as vasopressin). At the end of his 1906 paper, his only reference to effects on the uterus is amongst "incidental observations," where he concludes, "The pressor principle of the pituitary (infundibular portion) acts on some constituent of the plain muscle fibre other than that which is excited by adrenaline and by impulses reaching sympathetic axon-endings."[5]

In 1909, Dale published a slightly extended account of his experiments on cats using a more specific extract (of the posterior pituitary alone).[6] However, studies on the uterus occupy just one page, a page notable for a telling mistake. In that same year, Blair-Bell and Pantland Hick had shown that the extract could stimulate uterine contractions in rabbits and, importantly, they noted that the effects were stronger in pregnant rabbits than in virgin rabbits.[7] Dale denied that the effect was any stronger in pregnancy, conforming to his belief that it reflected a general action on plain muscle and explaining his lack of interest in the uterine actions. This was a mistake: in all mammals, uterine sensitivity to oxytocin increases markedly towards the end of pregnancy. Had Dale recognized this, he would have recognized that the oxytocic effect could not be an instance of the already known pressor response.

Thus, one of the most commonly stated "facts" about oxytocin—that "oxytocin was discovered by Henry Dale in 1906"—bears little resemblance to anything observed by Dale and none to anything claimed by Dale.[8] As a "fact," this statement is entwined with interpretation, and it makes no sense in historical context. It is this type of error that Kuhn had sought to avoid in his book, *The Structure of Scientific Revolutions*.

This is an error that assigns precedence to an observation rather than to an idea. All observations are bound up with ideas about what they mean. Dale's observation was bound up with an idea that was ultimately refuted—that the oxytocic effect of the pituitary extract reflected the actions of a pressor principle. Blair-Bell, on the other hand, formed the idea that a principle secreted from the pituitary controlled parturition—an idea that he followed up by testing the extract in pregnant women.

If the oft-repeated "fact" that oxytocin was discovered by Henry Dale in 1906 is no such thing, why has this not been noticed? The obvious answer is that those who cited Dale (1906) as the discoverer of oxytocin had not read Dale (1906). Does it matter that they mis-cited Dale (1906)? Not really—why should it matter who discovered oxytocin? So if it doesn't matter who discovered oxytocin, why do they cite Dale (1906) when they haven't even read it? The obvious answer is that to do so advertises the apparent erudition of the author. It is a rhetorical device, to persuade the reader that the author is deeply knowledgeable and attentive to the history of the topic. Dale (1906) has become a *totem* of the field.

What *Is* Oxytocin?

As we have seen, oxytocin has many identities: it has been an extract, then a chemical, a hormone, a neurotransmitter, a neuropeptide, the hormone of childbirth, the love hormone, the cuddle hormone.

It is not just the popular understanding of what oxytocin *is* that has changed. Nor is it just that oxytocin's identity depends on the focus of a particular author. Its technical meaning has also changed.

For Kamm, who named it, oxytocin was the active principle in an extract of the posterior pituitary, defined by certain biological effects. The widespread use of this extract in obstetrics made it necessary to establish a "standardized" extract, and in 1925 a committee of the League of Nations proposed that the *United States Pharmacopeia* standard reference powder be adopted. This was made from the posterior pituitaries of cattle, and 0.5 mg of powder was declared equivalent to 1 IU of the extract. For du Vigneaud, however, oxytocin was a sequence of amino acids, postulated to be identical with Kamm's principle. Of the 0.5 mg of powder that contained 1 IU of oxytocin, just 2 µg was oxytocin itself, as identified by du Vigneaud.[9]

Oxytocin is made in the brain as part of a large precursor molecule that is cleaved by enzymes to produce the active molecule. So if, as happens, enzymes cleave the precursor to yield a slightly longer or shorter version of oxytocin, is this still "oxytocin"? The answer might depend on who you ask—a chemist might repeat that these are longer and shorter versions of oxytocin. A physiologist though, might say that it depends on whether the different versions retain the biological activity of oxytocin.

When oxytocin is released into the blood it is degraded by enzymes. However, blood is full of large and sticky molecules—immunoglobulins—and some oxytocin gets tangled up in them. Bound oxytocin has no biological activity, because to affect any tissues in the body it must leave the blood to enter the fluid that bathes those tissues. Bound oxytocin accumulates in the blood, to levels that can greatly exceed those of "free" oxytocin: it is eliminated only when the molecules that bind it are themselves eliminated, and they have a lifetime of days whereas free oxytocin has a lifetime of a few minutes. So is this still "oxytocin"? A physiologist might say, just as oxytocin ceases to be oxytocin when it is inactivated by enzymes, so it ceases to be oxytocin when it is inactivated by binding. In this, they would be following Kamm's definition of oxytocin by its biological activity, and staying true to the meaning of the international units by which its levels are still reported.

For those who measured oxytocin by some assay, it was something recognized by that assay, assumed to be the same as that chemical sequence. However, the assays by which oxytocin is now measured are not measures of biological activity, and, depending on the assay, they might or might not "see" the bound oxytocin. Is an assay of bound oxytocin a meaningful measure, either of the production of oxytocin or of its activity? Probably not, because every individual has a different complement of immunoglobulins, a complement that changes with physiological state and life experience. Measures of bound oxytocin reflect the abundance of particular immunoglobulins at a particular time. So, if we measure bound oxytocin, we are probably not measuring anything relevant to the production, secretion, or actions of oxytocin.

While most mammals make oxytocin, marsupials make mesotocin—a molecule that differs from oxytocin by just one amino acid. This is a neutral mutation: mesotocin acts on the same receptor as oxytocin and has identical biological actions. Like oxytocin, mesotocin in marsupials is made in the hypothalamus, is secreted from the posterior pituitary gland, and regulates milk-ejection and parturition. So, is *this* oxytocin?

A chemist would say no; mesotocin is a homolog of oxytocin. But we take a different approach with other hormones, like prolactin, growth hormone, and insulin. Most of these vary, sometimes considerably, from one mammalian species to another, yet we call them by the same name—and why not?—a spade is a spade by virtue of its function not its shape.

In fish, another homolog of oxytocin, isotocin, is secreted from the posterior pituitary; it regulates egg-laying and has other actions akin to the effects of oxytocin on behavior, energy balance, and electrolyte balance. Some scientists call *this* oxyto-cin, rather than isotocin, because they consider themselves to be studying the same hormonal system, albeit in a different species, and because they want their work to be recognized by scientists who work on humans and other mammals.

But surely, if we confine ourselves to studies that have mea-sured oxytocin in human plasma, and only recent studies using widely used techniques, then we can be confident that these studies are talking about the same thing. Can't we?

In 2015, one of the largest studies of oxytocin in humans reported that the average plasma levels in 552 men were 0.4 pg/ml.[10] It used a radioimmunoassay that was "well validated" as producing measurements consistent with measures of biological activity; the methods involved "extracting" samples to eliminate large molecules that affected the assay performance. The levels reported were broadly consistent with many others using similar methods.

In 2013, a similarly large study reported average levels of 400 pg/ml in men.[11] This used a different assay—an enzyme-linked immunoassay—and the authors did not extract the sam-ples. Again, the levels reported were consistent with many other reports using that same method.

There is a 1,000-fold discrepancy in the reported levels between these two studies. Four published studies measured the same samples using these two methods. Each reported no cor-relation between the results of the two assays. Whatever it is that is measured in these two assays, it is not the same thing. It seems absurd that two contemporaneous communities of researchers, both working on oxytocin, should each be generating hundreds of papers nominally about oxytocin, but reporting on things that are completely uncorrelated.[12] What is going on?

Clearly these communities are not interacting effectively. One uses radioimmunoassays, the technique that was developed in the 1970s and maintained by specialized laboratories. The methods involve "extracting" samples to eliminate large molecules that affect the assay. However, such assays are time-consuming, and they depend on access to sensitive antibodies raised in animals, and stocks of these are diminishing; raising antibodies and refining and validating assays all require a particular expertise. The generation of scientists that developed them is retiring or has retired.

The other community comprises a new generation of scientists looking for correlations between oxytocin and psychological features in humans. It has mainly used commercial assay kits to measure oxytocin, kits designed to measure oxytocin in vitro not in blood, and they make no effort to extract the samples to remove interfering factors. Understanding of the necessity of this has been lost, and the discrepancies with the older assays are not noticed or are ignored because the community of psychologists has little interaction with the community of endocrinologists.

Because papers are reviewed by scientists within the same microcommunity, issues of wider external validity are not addressed. Techniques are assumed to be valid because they have been used by other members of the community, not because they have been objectively validated.

Textbook Facts

So what facts are absolute and immutable products of the progress of science? Shall we find them in textbooks of physiology? There, we find a curiously consistent, sadly atrophied, oddly selected account. One of the most commonly expressed facts about oxytocin in this landscape is that it is produced in the paraventricular nucleus of the hypothalamus, whereas vasopressin is produced in the supraoptic nucleus. This is a wave at the old, simplistic idea that regions of the brain can be classified by function. The development of immunohistochemistry enabled peptides to be "seen" in cells, and, more than fifty years ago, it allowed oxytocin to be seen. The earliest such studies and every study since showed that both the supraoptic nucleus and the paraventricular nucleus contain both oxytocin and vasopressin neurons. These together contain only about half of the oxytocin neurons—the rest are scattered across the anterior hypothalamus.

Some textbooks qualify their statement. They do so with words, still wrong, that *most* oxytocin neurons are in the paraventricular nucleus, that a *few* of each type are present in the other nucleus. Some statements are oddly specific—about a sixth of the neurons are in the other nucleus. This is repeated in several textbooks; it comes from no published analysis that we have found but is contradicted by many.

Perhaps we are pedantic: the textbooks are simplifying and the details don't matter. But that is our point—it doesn't matter that this "fact" is wrong, because it doesn't matter to anyone who will read the textbooks. It might help them to know that oxytocin is made by neurons in the hypothalamus and not in the pituitary gland, but why do they need to know details which, even if accurate, would have no significance for them and no connection with anything else in the textbook?

The first answer is that textbooks have their own evolutionary lineage, they evolve with minor mutations from older textbooks, and the many "facts" that don't matter go uncorrected because there is little selection pressure against them. They remain, pseudofacts in the junk of the textbook genome.

The second answer is that such facts, in being apparently precise and specific, are more memorable than the fuzzy, messy, qualified reality. The memory of them is more easily assessed by teachers who are assessing not knowledge or understanding, but mere engagement with the textbook. In an alternate world, textbooks might offer a "taster" of facts, as chocolatiers present a selection of truffles, each accompanied by notes that describe their loving and careful construction, presented on a silver plate that has its own story. Instead, many textbooks scatter facts like jellybeans on a plastic plate. In this alternate world, the student would learn that the "facts" as presented are each the unique expression of the scientist's art; they are exemplars of a process, a process that consists of observations, experiments, theory, and interpretation. They are exemplars whose life does not persist beyond the act of consuming them, but whose taste might linger and be remembered. That memory might inspire the readers to be chocolatiers themselves, or at least to respect the skill and wit that went into their construction—and understand the boundaries of uncertainty, and the changing nature of "facts."

The Social Construction of Facts

Each statement in a scientific paper about evidence requires interpretation and evaluation. It is addressed not to the casual reader, but to experts who share a common understanding of the terms used, and a common appreciation of the significance of the statements. How those statements are interpreted depends on *who* is doing the interpretation (with what technical knowledge of the methods used), *when* (with what knowledge of other "facts" in the literature), and *why* (exactly what in the statement matters to the reader). How they are evaluated involves weighing the strength of the evidence. This involves assessing not just the internal consistency of a paper—its methodological validity and analytical rigor—but also its external consistency with other published work.

We began the last chapter with what oxytocin is—a hormone, something secreted from cells in one part of the body and carried by the blood to act on cells in another part—and that it is secreted from the posterior pituitary gland. We now see that these things have little to do with what most scientists who work on oxytocin think of as oxytocin. The notion of a hormone has changed to encompass many things not carried by the blood and some that act very locally. The apparent roles of oxytocin that attract most attention are of oxytocin released within the brain, not into the blood, and the statement that it is secreted from the posterior pituitary is only true for all vertebrates if we count some things as oxytocin that were once considered different hormones.

Inescapably, "oxytocin" is a social construct. Its meaning depends on who uses it, who the intended audience is, and on accepted conventions and the particular time and context in which they apply. This is not to suggest that what we say of oxytocin is not anchored to the real world—just as statements in the scientific literature are facts that anyone can confirm by a visit to a library, so statements about experimental findings can be confirmed (or contradicted) by repeating those experiments. But how we interpret those findings, how we evaluate the strength of evidence that they embody, and how we assess their importance, all depend on social conventions. This is also true of how definitions of science are constructed—definitions reflect what they are used for by scientists. This does not mean that they are detached from the empirical world or that they are infinitely malleable, but rather that, because reality is complex, the way we partition it into categories requires choices, choices guided by what we are trying to achieve and by what we already understand.

Weighing and interpreting evidence requires scholarship, the process of rigorous inquiry that advances the teaching, research, and practice of a given academic field. It requires understanding the methods used, an appreciation of their strengths and limitations and an appreciation of what is involved in a systematic search and meta-analysis. It also requires philosophical as well as technical understanding of the methods, and an awareness of the social context of the science and of how this impacts its evolution and its perceived importance. Subsequent chapters engage with features of the scientific literature that require particular awareness—including publication bias, selection bias, confirmation bias, and some generic problems in science, including common issues in statistical analysis.

Postscript

Owain was just five when he asked his father for the first time about sex. This moment came sooner than expected. His father could safely skip the structure of steroid receptors, but where should he start? What could he assume that Owain knew? Should he give just the facts, all the facts he knew, and let Owain make sense of them? Should he impose some structure on the facts, and if so what structure? There would be no gooseberry bushes and no moralistic homilies. He would suppress what he knew of the correlation between the birthrate in Western Europe and the stork population.[13]

But if he would give just some of the facts, then which facts? Did he know any facts anyway? In Karl Popper's words, "Science does not rest upon solid bedrock. The bold structure of its theories rises, as it were, above a swamp."[14] We must choose which facts to tell, and our choices betray our prejudices and preconceptions, and we will choose those facts knowing that few of them are unchallengeable.

So what should Owain's father say to Owain? Does he tell it all: all the detail, the confusion, the messiness, the doubts, the ambiguity? He must ask a question first, the most important question always—"Why do you want to know?" This question, a question that we must ask ourselves again and again, defines the level of detail and explanation that is appropriate.

What was Owain's answer? He needed to know for his homework. What, exactly, did he need to know? Well, he had this form to fill in. What sort of form? One with a box in the top corner that said Sex: M/F.

This, his father could help with.

15 Organized Skepticism in Science

It is a part of the business of the critic to preserve tradition—where a good tradition exists. It is part of his business to see literature steadily and to see it whole; and this is eminently to see it as not as consecrated by time, but to see it beyond time . . .

—T. S. Eliot, quoted by E. M. Forster in *Aspects of the Novel*[1]

In chapter 13 we presented the history of research in oxytocin as observed in the topics of the papers that had seemed to be most influential. We hinted that some of these were controversial, that the work on maternal behavior, for instance, had been challenged and had survived that challenge. We mentioned that a large body of work measuring oxytocin has been challenged on the basis that the assays were unreliable. We hinted at uncertainty about how much oxytocin enters the brain after intranasal administration.

Any close reading of any scientific literature in any period will find contradictions, revisions, qualifications, and reinterpretations. Mostly, these are the "noise" of experimental uncertainty and methodological variability, with little lasting consequence for scientific understanding. The oxytocin field has not been injured by fraud in ways that have been noticed, nor is it a field where conflicts of interest appear prominent. Occasional papers have reported unbelievable results; the fate of these, by and large, has been to be ignored.

In the oxytocin field, research activity accelerated massively following a paper that reported effects of intranasal oxytocin on trust in humans. But there is skepticism: in 2015, a meta-analysis of studies concluded that "the simplest promising finding associating intranasal [oxytocin] with higher trust has not replicated well."[2] It noted that other evidence claimed as circumstantial support for the trust hypothesis was flawed: evidence from measuring oxytocin in plasma came from unreliable assays, and while some small studies had reported evidence of associations of specific [oxytocin]-related genetic polymorphisms and trust, larger studies had not replicated these findings. The authors concluded that "the cumulative evidence does not provide robust convergent evidence that human trust is reliably associated with [oxytocin] (or caused by it)."[3]

Several other reviews have criticized studies of intranasal oxytocin. One concluded that "intranasal [oxytocin] studies are generally underpowered and that there is a high probability that most of the published intranasal [oxytocin] findings do not represent true effects."[4] Another argued that very little of the huge amounts applied intranasally reach the brain, but that peripheral concentrations are raised to supraphysiological levels, with expected effects on diverse targets including the gastrointestinal tract, heart, and reproductive tract.[5]

Skepticism might be defined as doubt, not denial, and is expressed as a call for more, or better, or different evidence to support some contested interpretation, or for an explanation of internal or external inconsistencies in the evidence. If scientists are passionate in their claims, there is passion also amongst the skeptics. It is in the critical attitude that we see the force of Merton's norms, and their ability to improve Science. There are few Damascus moments in Science; those who have fortified positions do not surrender them lightly. Yet where there is debate, there are eyes to read it and ears to hear it, and the uncommitted, the students, the newcomers, pay close attention, for the decision of whether to invest their futures in a field is not one to be taken lightly.

Concerns about replicability and rigor wrack many fields of science. Yet it seems obvious that oxytocin research over the last hundred years constitutes progress, even if errors and flaws are abundant. For scientists, this is what progress typically looks like: a steady expansion of the depth and the breadth of understanding. They have learnt ever more about what oxytocin does (through observational studies), ever more about how that knowledge might be applied (through intervention studies), and have built a detailed understanding of the fundamental mechanisms. They have developed a theoretical vision that appears to unite the totality of evidence.

How has this been achieved? A conventional view, one shared by many scientists, historians, and philosophers, is that the credit lies with the scientific method: that the self-correcting nature of science draws its products ever closer to objective truth.

This is an appealing narrative—simple, memorable, and reassuring—but there is a circularity: to claim that science is self-correcting is to assume that current notions are correct and that those that have been superseded are wrong.

The main self-correcting mechanism has been understood to be the processes of replication and refutation, by which errors are detected and eliminated. But it is rare that experiments are directly replicated, or at least, rare that replications appear in the scientific literature.[6] Journals don't like to publish direct replication studies—they favor originality. Nor have funding bodies favored them. Instead, most replication studies are "conceptual" replication studies—they do not reproduce the design of the original study, but find a new way of testing the conclusion. The problem is that when conceptual replications fail to produce the expected outcome, this failure can easily be dismissed on the basis of the methodological differences.

Scientists who seek to extend the work of others will often begin by reproducing previous findings, and may mention these efforts in their papers. But if they fail to reproduce earlier findings, their natural response may be to move to another question. The failure to reproduce an outcome might speak of many things—of technical competence, or of unconsidered confounds in the experimental conditions—and excluding these possibilities can be a long commitment with the prospect of little return. Moreover, as shown in chapter 4, the refutation of a hypothesis is not a cause for celebration, but for disappointment. The papers that destroy a hypothesis are not likely to be cited.

Thus, there is there little joy to be had in destroying a cherished idea. However, this seldom arises: to destroy a cherished idea takes exceptional effort once it has taken root. Once an idea has a broad base of empirical support, has been endorsed by numerous authorities, and has found its way into reviews with "fact-like" status, it can be very hard to dislodge. Strong structural forces lead to the dogmatic persistence of some notions and to the neglect of inconvenient evidence.

The Conservatism of Science

First, there is a "founder effect." It is easier to publish findings that support an established position than findings that are inconsistent with it. Because scientists generally have flexibility in how they interpret evidence, they will often use that flexibility to avoid conflict.

Second, there is "confirmation bias." When analyzing evidence, scientists interpret it in terms of what they expected to find—and what they expect to find, is, generally, what other scientists have found.

Third, there is a "file drawer problem": scientists don't bother publishing negative results. Journals are unenthusiastic about negative studies, and referees set them a disproportionately high bar. This means that the literature gives a misleading account of the totality of evidence.

Fourth, negative studies that are published are cited less often than positive studies. Even if the published evidence is mixed, one side gets less attention.

However, some ideas are overthrown, as in the following case study.[7]

The Vasopressin–Memory Hypothesis: Analysis of a Debate

As described in chapter 5, hormone secretion from the pituitary is governed by the brain. When this idea became established, it led to speculation that these hormones might act back on the brain to affect behavior. In the 1960s, David de Wied focused on the idea that the hormone *vasopressin* was involved in memory.

The main function of vasopressin is on the kidneys: it is secreted in response to dehydration, and its effects on the kidneys minimize the water loss that accompanies the production of urine. But it is also secreted in stressful conditions, and it has the "pressor" effect for which it was named—it raises blood pressure. De Wied pursued the idea that vasopressin might influence how well a stressful experience was remembered.

In his experiments, rats learnt that the sound of a buzzer was a warning that an electric shock would follow, and they learnt to avoid this by crossing a barrier. De Wied studied how quickly rats learned this, for how long they remembered it, and he tested whether either the learning or the memory were influenced by vasopressin. He found that injections of vasopressin could facilitate this memory. Importantly, he gave the vasopressin not into the brain but by injections under the skin, and the effectiveness of this "subcutaneous" application was key to the idea that vasopressin secreted from the pituitary acts back on the brain.

Throughout the 1970s, de Wied's studies, developed with a network of collaborators, attracted increasing interest. They were published in leading journals, and from the mid-1970s, reports began to appear that intranasal vasopressin had beneficial effects on memory in humans.

The core experimental foundations of de Wied's hypothesis combined four claims:

1. The effects of vasopressin in certain behavioral tasks reflected effects on the consolidation and retrieval of memories.
2. Vasopressin had these effects by virtue of its actions on the brain.
3. Vasopressin secreted from the pituitary was important for memory.
4. Vasopressin injections could reverse induced memory deficits and might be of therapeutic value in humans.

Each of these claims was supported by evidence, but not everyone was convinced. An important part of the case had come from experiments on Brattleboro rats—a species that, because of an inherited genetic defect, cannot produce vasopressin. These rats, according to de Wied, had poor memory, and this could be corrected by giving them vasopressin. But not all groups could find the defects. It seemed either that the Brattleboro rats raised in different laboratories differed in some unknown ways, or else that there were differences in how memory was measured in the different labs.

De Wied had interpreted the outcomes of his studies as evidence of effects on memory, but some who entered the field became disillusioned. They began to think that anything which affected attention, arousal, or emotionality might have a similar effect. These critics went on to show that vasopressin indeed seemed to have nonspecific effects. They showed that subcutaneous injections of the large doses of vasopressin that de Wied had used produced large increases in blood pressure, and that rats found this unpleasant. They showed that vasopressin did indeed affect arousal in rats and that any effects on arousal produced similar effects on the memory tests.

Moreover, there was skepticism about whether enough endogenous vasopressin would enter the brain to have any effect there. A blood–brain barrier protects the brain from large molecules that circulate in the blood. This, it was argued, should prevent the entry of vasopressin.

De Wied saw that this was a serious objection, and a PhD student, Wim Mens, was set the task of measuring exactly how much of the vasopressin given in a peripheral injection entered the brain. In 1983 the results were reported by Mens and colleagues.[8] The last sentence of the abstract states, "The present results demonstrate that neurohypophysial hormones do cross the blood–brain barrier in amounts obviously sufficient to induce central actions." But this claim sat uncomfortable after the preceding statement, which declared the actual extent of passage: "approximately 0.002% of the peripherally applied amount [. . .] reached the central nervous system at [ten minutes] after injection."

It seems that, before doing these experiments, the authors were certain that, for vasopressin to have effects on memory, it had to have entered the brain. Therefore any amount of passage into the brain, however small, had to be enough. The authors were trapped by this logic. When the results showed a miniscule amount of passage, they were forced to either confront a contradiction with the de Wied canon or else to use "flexibility of interpretation" to an absurd degree. They chose the latter.

As the case for vasopressin crossing the blood–brain barrier in sufficient quantities looked increasingly unlikely, the hypothesis was reformulated. With the advent of immunocytochemistry, vasopressin was found in cells and in nerve fibers. This could not explain how peripherally injected vasopressin affected memory, but it offered a way of saving the idea that vasopressin might have some role in memory.

This idea foundered as the diversity of vasopressin pathways came to be recognized. Vasopressin was found in neurons that control circadian rhythms of behavior, and in neurons implicated in temperature regulation, aggressive behavior, and cardiovascular regulation. Each of these discoveries gave reasons to think that a behavioral test which appeared to be measuring memory might be measuring nothing of the sort.

The only point of rescue appeared to be evidence that vasopressin might improve memory in patients. This had come from studies using intranasal vasopressin, and the first two, published in the *Lancet*, attracted considerable attention despite being very small. In one, cited 226 times, twelve patients given vasopressin performed better in memory tests than eleven given a placebo. In the other, cited 175 times, four patients with amnesia were reported to have improved after being given vasopressin; there were no controls for this study. These were swiftly followed by two negative studies also published in the *Lancet*, but which were cited much less often.

By 1992, forty-four studies of intranasal vasopressin had been published. Eighteen were consistent with de Wied's hypothesis, but twenty-six were not. The supportive studies have been cited on average seventy-six times and the negative studies just twenty-four times. Nevertheless, the discordance led this line of investigation to be abandoned.

By the end of the 1980s the hypothesis was in disarray. The studies on Brattleboro rats seemed to imply that complete absence of vasopressin had no clear consequences for cognitive function. Vasopressin penetrated the brain in, at most, tiny amounts, and it seemed that the behavioral effects could be attributed to effects of the resulting increase in blood pressure. Given the evidence of diverse central sites of action of vasopressin, the possibility that the apparent effects on memory reflected incidental actions in the brain seemed impossible to exclude. Prospects of a therapy for memory deficits in humans, despite the selective citation of positive results, had dissolved.

The hypothesis had rested on several lines of evidence. An attack on any one alone could not have been decisive; failed replications are seldom damning, because there is always flexibility of interpretation.[9] But, in the early 1980s, each pillar was the subject of assault by several independent skeptics, and the critical strands were drawn together to make a comprehensive case against the hypothesis.

The controversy is now largely forgotten, and both the papers that laid the foundations of the hypothesis and the papers that dismantled it are now seldom cited. The debate went largely

unnoticed by those not directly involved; there was no final declaration of the death of the hypothesis; it languishes as a pseudo-
fact in the genome of the vasopressin field. News of its failings
spread more by word of mouth, in conferences and discussions
between scientists; the critics left the battle, their work done. Yet
reviews written by scientists distant from the controversy still
often cite effects on memory as an accepted part of the canon.
There is little comfort for those who hold that science is self-
correcting in this.

However, de Wied's work stimulated interest in the central
release of vasopressin. De Wied had coined the term "neuropeptide" to describe vasopressin's role in information processing.
He had recognized that these actions were conceptually distinct from what was understood to be the role of conventional
neurotransmitters, and this prompted feverish interest in other
neuropeptides, leading to fundamental changes in our understanding of information processing in the brain. Today we recognize about three hundred neuropeptides made throughout the
brain, messengers important in a vast range of behaviors and
physiological functions, and drugs based on neuropeptides have
been developed for a host of neurological disorders. De Wied's
hypothesis was massively important despite being wrong.

In this story, we might see a different understanding of progress in science. The idea that science advances by continual
introspection seems to conceive of a scientific field as constituting scientists driven by common goals, that there is a purpose
behind their activity, and that the intent can be recognized in
retrospect by what has apparently been achieved by that activity. This neglects the most striking characteristic of science
observed by Price—its continuing exponential growth. So far
from being introspective, science is exuberantly extroverted—its

canvas is continually expanding at all places in all directions and in all available ways. Where these paths lead is inherently unpredictable.

The idea that because a scientific theory is false means that research on it has been wasted is questionable. De Wied's theory was wrong, yet led to a radical change in our understanding of how neurons in the brain communicate. Probably the paper on intranasal oxytocin and trust is wrong. Few now take seriously that oxytocin is involved in trust—but whether the findings are right or wrong are beside the point. The paper continues to be cited because it transformed the field by illustrating a novel methodological approach.

Talking of waste in these terms doesn't make sense. You might say that a study that confirms something known or a study that finds nothing is wasted effort. But if a finding matters, it needs to be checked, and scientists have to look before they can know that there's nothing to see. Scientists will make many mistakes when they try to be bold. But bold ideas invigorate the community, feed the passion that makes science grow—and with that growth will come breakthroughs, will come transformations—where, we can't know, but they will come somewhere, and they will come from someone passionate, and they might even come from mistakes.

Postscript

The Mens et al. paper is now a classic for its elegant quantification of the efficacy of the blood–brain barrier, and its results have since been replicated in other species. However, it was never cited by de Wied or his collaborators. But, to crib a term used in scientometrics, Mens et al. was a "sleeping beauty." Most papers are cited most in the first few years after publication and then slowly forgotten. Mens et al. was cited on average eight times per year in its first eight years, "slept" for nineteen years, with an average citation rate of three per year, and "reawakened" in 2013, with an average of sixteen citations per year in the next six years. Its current higher rate of citation reflects its prominence as a focus for doubts about whether enough intranasal oxytocin reaches the brain to explain its reported effects.

There are much more spectacular sleeping beauties, papers that display the long lags that can separate an idea and its practical translation. As we highlight in chapter 18, most of the one hundred most-cited papers of all time as reported by *Nature* in 2014 are methods papers.[10] But the list includes a few exceptions. One, sneaking into the list at number 99, is a paper by Edward Teller, later known as the "father of the hydrogen bomb." It presented what is known now as the Brunauer–Emmett–Teller (BET) theory, a theory that is the basis for measurements of the specific surface area of materials.

Teller was thirty years old when he published this paper in 1938. It was cited just seventeen times in the next five years, but from 1994 on, the citations took off (figure 15.1). Its impact had begun after a lag of fifty-six years.

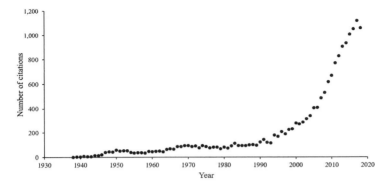

Figure 15.1
Citations per year to S. Brunauer, P. H. Emmett, and E. Teller, "Adsorption of Gases in Multimolecular Layers," *Journal of the American Chemical Society* 60 (1938): 309–319.

16   Webs of Belief: Citation Networks

How and why does knowledge circulate? How does it cease to be the exclusive property of a single individual or group and become part of the taken-for-granted understanding of much wider groups of people?

—James A. Secord[1]

Scientific knowledge is deposited in scientific papers, and what scientists know comes mainly from reading them, or from hearing about them at conferences. Each non-novel claim that is repeated in them (and which has not yet achieved "fact" status) is expected to be referenced in a way that directs the reader to evidence for that claim, either to primary studies or to reviews that summarize the findings of primary research.

The scientific literature in any given field is a network, with connections from each paper to their cited papers forming a web of knowledge. To map the relationships between papers, *network analysis* is particularly suitable. The mathematics behind network analysis is complex, but the underlying logic is not.

Figure 16.1 shows a schematic of a small citation network; the circles represent papers, the links between them are represented

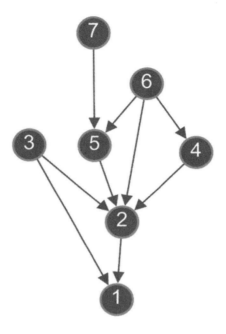

Figure 16.1
Citation Network A (n = 7 papers; m = 9 citations)

by arrows that also convey information about the direction of the relationship.

Price used such analysis to generate the findings described previously, findings that suggest that the system of scientific communication is governed by a dramatic inequality: a small number of publications attract the lion's share of citations.

The convention of citation and network analysis provides an opportunity to study the diffusion of ideas, and in 2009 Steven Greenberg, a neurologist at Brigham and Women's Hospital in Boston, did just this.[2]

Claim-Specific Networks

Greenberg became interested in a specific claim: that a particular protein, β-amyloid, was abnormally present in the muscles of patients with *inclusion body myositis*, a muscle-wasting disease. This claim had important implications for treatment strategies, and Greenberg had seen it repeated in at least two hundred papers that had given the impression that this was a "fact." He wanted to find the evidence for it.

After searching the literature, Greenberg found just twelve papers that had investigated the claim directly. Of these, six supported it, but six did not. To his eyes, when considered together, the evidence was very weak. Worryingly, the first four supportive papers all came from the same laboratory, and two of these "probably reported mostly the same data without citing each other."[3] On his reading of these papers there were major technical weaknesses, "most notably a lack of quantitative data as to how many muscle fibers were affected and a lack of specificity of reagents."[4]

Greenberg wanted to understand why this claim, that to him appeared questionable, had become an apparent "fact." The first ten of the primary studies were all published between 1992 and 1995, and he looked at how these were cited in the years until 2007. He found 242 papers that discussed the presence of β-amyloid in patients with inclusion body myositis. He recorded all 675 citations from one paper to another in this set to understand the interactions between them. He then classified every citation as being either: (1) supportive of the claim, (2) neutral, or (3) critical of it, to construct a *claim-specific network*.

The network looked like a typical citation network—it had a power-law-degree distribution of citations. A total of 214 citations were directed to the early primary studies, but 94 percent of these were to the four supportive studies, and just 6 percent to the six unsupportive studies. The literature was overwhelmingly citing supportive evidence while neglecting unsupportive evidence—a phenomenon known as *citation bias*. But how did this happen?

Greenberg analyzed how information spread through the network. Citation networks had been used previously to understand the intellectual structure of disciplines and fields by grouping papers together that shared references or citations. They had also been used to understand how research fields progressed by establishing the relationships between important contributions to the literature. They had provided information scientists with data that they could compare to models of citation behavior— such as Price's cumulative advantage model. But Greenberg did something novel. By classifying both the papers and their citations by their position on this claim, he traced the propagation of bias and its influence on opinion.

Greenberg looked at the top ten most highly cited papers in his network. The four supportive primary studies were all among the top ten—and these four were all written by the same research group. The other top-cited papers included five studies in animal models of the disease, and one review paper, and all of these expressed the view that the β-amyloid claim was true. These did not contain relevant primary data, and each of them cited only supportive primary studies.

Citations help to understand how documents relate directly together, but Greenberg also measured the number of *paths* in the network to understand how information might have spread. In this context, a path is a chain of citations linking a series of papers together. For example, if *A* cites *B*, which cites *C*, there is a path from *A* to *C* via *B*, and this helps us trace the provenance of an idea.

By counting paths and the papers through which they passed, Greenberg showed that review papers had played a major role in directing scientists to the evidence. Knowledge about β-amyloid derived from four reviews all written by one group of scientists: 95 percent of all paths to the original primary data went through these papers. These reviews cited the four supportive primary papers but none of the unsupportive studies—they funneled attention to studies that supported the β-amyloid claim.

> Between 1996 and 2007 support for the claim grew exponentially, with the number of supportive citations and citation paths increasing sevenfold and 777-fold, to 636 citations and 220,553 citation paths. In contrast, the critical view grew to only 21 citations and 28 citation paths [. . .]. No papers refuted or critiqued the critical data, but instead the data were just ignored.[5]

He also found evidence that claims that began as conjectures were being converted into "facts" in the literature by a process of citation alone. One claim, repeated in twenty-four papers, was that the accumulation of β-amyloid in muscle fibers precedes other abnormalities. It had begun as a hypothesis, but some papers claimed that it was probably true, and eventually papers were stating it as a fact. But no data directly relevant to this had been published; papers were citing other papers that simply asserted it. Through chains of citations linking papers containing no relevant data this claim became a "fact."

Extending his study to nine funded research proposals, Greenberg found citation bias and other citation distortions there, too. In six of the nine proposals, only supportive primary data were cited. In two of the three proposals that cited at least one unsupportive primary study the meaning of that data had been changed to make it appear supportive. A biased and distorted presentation of evidence was being used to win research funding.

Greenberg's findings suggested that the scientific community had accepted the claim because they only saw the supportive evidence. Their beliefs appeared to have been guided by reviews that ignored contradictory evidence. Belief was further amplified by reviews that erroneously interpreted the contradictory evidence. These had provoked a cascade of information in which papers repeated the erroneous conclusions of those reviews rather than studying the primary literature. The "shape" of the literature, as connected through citations, thus affects both what evidence is used and how it is interpreted.

These are worrying findings. Next, we look at two studies that used similar methods that suggest that such citation bias has infected the evidence base in areas of particular importance for public health.

Citation Bias and Distortions

Often, the literature relevant to a research question will be too large for a paper to reference adequately and too time-consuming for scientists to evaluate. As a result, papers only reference a small, selected sample of the relevant literature. This sampling is not regulated by any formal convention except in the (rare) cases of "systematic reviews" of the literature. We might hope that the account of evidence should be balanced, but often it is not. Papers disproportionately cite supporting evidence while neglecting unsupportive evidence—a phenomenon known as *citation bias*.[6]

In a claim-specific network, *authorities* have a particular influence, the papers in a network that receive the most citations.

Authorities are, according to Greenberg, papers whose major claims have reached a level of "social consensus"—with citations operating as a form of social approval. However, when reviews selectively cite supporting evidence, and when authoritative papers cite such reviews but not the primary papers, there is a danger of "amplification" of a false claim. The claim is likely to be repeated by many papers, but with no transparent connection to the primary evidence.[7]

Citation bias and amplification are major sources of citation distortion, but Greenberg identified many other problems:

- *Citation diversion*: erroneous citations that distort the meaning of a study's findings. For example, if paper *A* references *B* as supporting claim *Y* when in fact *B*'s conclusion contradicted it.

- *Citation transmutation*: the conversion of a hypothesis into a general fact through citation alone. For example, if paper *A* cites *B*'s assertion that *Y* is true but fails to recognize the qualified nature of that assertion.

- *Back-door to knowledge*: citing conference abstracts and other literature with no published methods or data and which hasn't been subject to peer-review.

- *Dead-end citation*: where a paper is cited in support of a claim despite containing no relevant material. For example, if paper *A* cites *B* as supportive of a claim when *B* says nothing about that claim.

- *Title invention*: the stating of a specific claim in a title or abstract although the paper does not experimentally or theoretically address that claim. According to Greenberg, "this method of invention can be successful, because titles or abstracts are often read and their claims repeated without reading the body of the paper."[8]

Although scientists may try to cite in an unbiased manner, small choices made in reviews over what papers to cite can have a disproportionate impact on the future shape of the scientific literature, which, in turn, influences future citing behavior. Next, we look at two studies that suggest that such citation distortions have infected the evidence base in areas of particular importance for public health.

Citation Bias and Research Underutilization

In 2018, a study used claim-specific citation network analysis to understand the propagation of evidence for the hypothesis that dietary saturated fats play a causal role in the development of coronary heart disease.[9] Between 1965 and 1968, four randomized controlled trials were conducted to test whether diets with a modified fat content would be beneficial for patients with preexisting heart disease. By 1984, these were still the only published randomized controlled trials to test this. In that year, a consensus conference of experts was convened which concluded that elevated serum cholesterol is a cause of coronary heart disease and that reducing fat consumption would be beneficial. The panel recommended that adults and children over two years of age eat a diet consisting of no more than 30 percent fat, reduce their intake of saturated fatty acids to less than 10 percent of total calories, and increase their intake of polyunsaturated fatty acids to 10 percent of total calories. Specifically, the panel stated that "studies are available that indicate a beneficial effect of treating high cholesterol levels in individuals with preexisting clinical disease."

As mentioned, these four trials were, at the time, the only direct evidence about the effectiveness of dietary interventions in such individuals. One of these trials, the Oslo study, had found lower levels of coronary heart disease in the intervention group, but the other three had returned equivocal or negative results.

Thus, while the Oslo study supported the case for dietary intervention, the other three were "unsupportive"—an evaluation supported by the directors of each. Taking the four trials together, there was no significant effect from lowering the fat content of the diet on the "hard" clinical end-points—the incidence of myocardial infarction, fatal heart disease, and all-cause mortality.

The three unsupportive trials were published in the *Lancet* and the *British Medical Journal*, while the supportive trial was published in *Acta Medica Scandinavica*. The three unsupportive trials were headed by leading authorities of cardiology, while the single supportive trial was a published doctoral thesis. Of these four trials, the Oslo study and the unsupportive trial of 1968 were the largest and longest, with similar sample numbers and protocols. It is hard to make a case that the Oslo study was more authoritative than the unsupportive trials.

Between 1969 and 1984, sixty-two review articles discussed the results of at least one of these trials. Of these, twenty-eight drew conclusions in support of dietary interventions, seventeen were neutral, and seventeen were unsupportive. But twenty-three of the supportive reviews had cited only the Oslo study. They did not critique the discrepant data; they either ignored it or were ignorant of its existence.

Polarization of Scientific Communities

In 2016, in "Why do we think we know what we know? A meta-knowledge analysis of the salt controversy," Ludovic Trinquart and colleagues asked whether differences in citation behavior could explain the ongoing controversy regarding guidance on the overconsumption of salt.[10] For one side of the debate, public health guidelines that encourage people to reduce their intake of salt are important to reduce the prevalence of cardiovascular diseases in the population as a whole. But for the other side of the debate, only some people (those with impaired kidney function) are at risk from excessive salt intake, and too little attention has been paid to the harms associated with low salt consumption in other subgroups.

Trinquart et al. identified 269 documents that examined the link between dietary sodium and either all-cause mortality or cardiovascular disease (CVD). They then classified each document by whether the major conclusion was supportive, inconclusive, or unsupportive of the hypothesis that salt reduction policies result in a reduction of either CVD or all-cause mortality. By their analysis, 54 percent of documents were supportive, 33 percent unsupportive, and 13 percent were inconclusive. They then constructed a citation network. Authors advancing a particular perspective were about 50 percent more likely to cite papers that agreed with their position. They concluded that

> The published literature bears little imprint of an ongoing controversy, but rather contains two almost distinct and disparate lines of scholarship, one supporting and one contradicting the hypothesis that salt reduction in populations will improve clinical outcomes.[11]

The opposing camps disproportionately cited data that supported their positions; they used different data, from different types of studies. Even in systematic reviews, there was uncertainty over what should count as evidence. The salt controversy was comprised of two groups of scientists arguing past each other because they drew from different sources of evidence that led them to reach starkly different conclusions.

Implications

One comfortable view of Science is that scientists conform to the norms of science, and evaluate all of the relevant literature and reference all of the work that is pertinent to their claims. Scientists, free of bias, evaluate studies by their methodological rigor, and come to objective conclusions. From time to time, scientists might disagree, but these disagreements are settled by new evidence or new theoretical explanations. In the absence of clear evidence, scientists acknowledge uncertainty. In these three examples, this picture of science is untenable. They show how long-term citation distortions undermine arguments that Science has an effective mechanism of self-correction.

In Greenberg's study, the four most highly cited supportive trials were produced by the same laboratory, and the four reviews that directed many later citations to these trials were also written by this group. These reviews failed to cite unsupportive trials, but this was not because they were unaware of them. They were aware of at least two of the unsupportive trials because they had conducted these themselves, but had nevertheless failed to cite them. It seems likely that they were influenced by confirmation bias—they weighed the supportive trials more favorably because they conformed to their expectations, and they dismissed findings that didn't fit. When later scientists saw only supportive evidence cited by reviews and by authorities, they didn't think to question whether unsupportive evidence existed. The architecture of the literature itself appears to influence what scientists cite. Scientists don't *defy* the norms of science—they just don't notice that they aren't conforming to them.

We might hope that the account of previous evidence in a scientific paper is balanced, but often it is not. Often, the literature relevant to a research question is too large for a paper to reference adequately and too time-consuming for scientists to evaluate. As a result, papers only reference a small, highly selected sample of the relevant literature. This sampling is not regulated by any formal convention except in the (rare) cases of "systematic reviews" of the literature.

Given continuing exponential growth in the scientific literature, it seems inevitable that scientists will increasingly depend on review articles as their means of recognizing the current state of knowledge. Formally structured meta-analyses, mainly in clinical medicine, follow guidelines that define a careful search strategy to ensure that their surveys of evidence are exhaustive, rigorous, and balanced. But for most reviews there is no such expectation. Reviews commonly draw not only from the primary literature, but also extensively from previous reviews. There are thus dangers that they introduce distorted representations of the evidence and that these will be amplified by the practices of citing reviews and of finding primary sources from their reference lists.

Our point in discussing these trials is not to claim that β-amyloid claim isn't true, that the overconsumption of salt is safe, or that altering the fat content of the diet isn't a sensible treatment for those recovering from myocardial infarction. Rather, these studies demonstrate that citation behavior can diverge starkly from an ideal—and what scientists read in papers cannot be taken at face value.

But how prevalent are these biases?

17 Unintended Consequences: Publication and Citation Bias

For as knowledges are now delivered, there is a kind of contract of error between the deliverer and the receiver: for he that delivereth knowledge desireth to deliver it in such form as may be best believed, and not as may be best examined; and he that receiveth knowledge desireth rather present satisfaction than expectant inquiry; and so rather not to doubt than not to err: glory making the author not to lay open his weakness, and sloth making the disciple not to know his strength.

—Francis Bacon[1]

Today, increasing attention is being paid to the quality of evidence used by politicians, the media, and scientists, and the manner in which evidence circulates is causing concern. Best known is the "fake news" phenomenon, where intentionally false stories, often originating from social media, online forums, and online "fake" news sources, seep into the mainstream media or into the rhetoric of politicians.

But some have begun inspecting the quality of evidence in the scientific literature, how that evidence is used, and how the spread of information shapes opinion in the scientific community. Scientists are particularly concerned about *dissemination biases*, some of which we touched on in the previous chapter. The existence of these is provoking concern that the integrity of Science is being undermined and that "myths" or "half-truths" can spread rapidly through the scientific literature.

Publication Bias

The best known of these biases is *publication bias*, which arises from scientists preferentially publishing experiments with statistically significant ("positive") outcomes. In 1979, psychologist Robert Rosenthal coined the term "file drawer problem":

> The "file drawer problem" is that the journals are filled with the 5% of the studies that show Type I errors, while the file drawers back at the lab are filled with the 95% of the studies that show nonsignificant (e.g., $p > .05$) results.[2]

By the late 1980s, publication bias had become known for its detrimental impact on medical advice. In 1987, Kay Dickersin and colleagues asked 318 authors of randomized controlled trials whether they had been involved in any trial that had gone unpublished.[3] The 156 authors who responded reported the existence of 271 unpublished trials—about 26 percent of all those they had been involved in. Of the unpublished trials, 178 had been completed and only 14 percent of these supported the theory under investigation compared to 55 percent of the published trials. It seemed that authors just didn't bother to write up negative trial results and submit them for publication. Since then, different forms of publication bias have been identified, including:

- *Time-lag bias*, where trials with impressive positive results (a large effect size, statistical significance) are published more quickly than trials with negative or equivocal results.

- *Outcome reporting bias*, reporting only statistically significant results or results that favor a particular claim while other outcomes have been measured but not reported.

- *Location bias*, publishing nonsignificant, equivocal, or unsupportive findings in journals of lesser prestige, while studies reporting positive, statistically significant findings tend to be submitted to better-known journals.

This problem has not gone away. In 2010, the US National Institute for Health Research conducted a systematic study of health-care intervention studies.[4] It found that studies with significant or positive results were more likely to be published than those with nonsignificant or negative results and tended to be published sooner. Published studies tended to report a greater treatment effect than studies in the so-called gray literature, which alludes to unpublished studies that have not been submitted to a peer-reviewed journal. It found that this bias affected consensus conclusions of systematic reviews.

In 2015, Michal Kicinski and colleagues examined 1,106 meta-analyses published by the Cochrane Collaboration on the efficacy or safety of particular treatments.[5] For meta-analyses that focused on efficacy, positive, significant trials were more likely to be included in the meta-analyses than other trials. Conversely, for meta-analyses that focused on safety, "Results providing no evidence of adverse effects had on average a 78 percent higher probability to enter the meta-analysis sample than statistically significant results demonstrating that adverse effects existed."

These were disquieting findings. Cochrane reviews are supposed to be gold standard in the biomedical area, but even here there was bias. The explanation might be either that (1) trials producing nonsignificant, equivocal, or unsupportive findings are failing to be published, or (2) they are being published but are ignored by meta-analyses.

Citation Bias

In the 1980s, researchers noted that studies reporting positive, statistically significant findings were cited more often than studies with equivocal or negative findings. The first systematic study of this was Peter Gøtzsche's 1987 study of clinical trials of anti-inflammatory drugs in the treatment of rheumatoid arthritis.[6] Gøtzsche looked at how the authors had referenced previous trials of the same drug. He searched the literature to find all published trials and classified each paper by whether their authors had interpreted the outcome with the drug as "positive" or "not-positive." He then looked for evidence of bias in the citations. Positive bias was judged to have occurred if the reference list disproportionately cited trials with positive outcomes. Of seventy-six papers in which such bias could potentially occur, forty-four showed a positive bias. Many authors had preferentially cited evidence that had shown a positive outcome for the drug that they were testing.

In 1992, Uffe Ravnskov looked at how trials of dietary interventions for coronary heart disease were cited.[7] Trials that supported their effectiveness were cited, on average, forty times per year, but unsupportive trials were cited just seven times per year. How often any particular trial was cited was not correlated with its size, nor with the quality of the journal that it was published in. Eight supportive trials were published in major journals and these had been cited on average of sixty-one times per year; ten unsupportive trials in similar journals had been cited just eight times per year. He concluded that: "the preventive effect of such treatment has been exaggerated by a tendency in trial reports, reviews, and other papers to cite supportive results only."

Citation bias is now a well-documented phenomenon. In 2012, Anne-Sophie Jannot and a team examined 242 meta-analyses published in the Cochrane Database of Systematic Reviews between January and March 2010, covering diverse research focuses, including cardiovascular disease, infectious disease, and psychiatry.[8] The 242 meta-analyses had referenced 470 unique trials. Trials with statistically significant results for the primary outcome accumulated, on average, more than twice as many citations as trials that did not show a statistically significant primary outcome.

In 2017, Bram Duyx and colleagues reviewed fifty-two studies of citation bias—thirty-eight on bias in the biomedical literature, seven in the social sciences, six in the natural sciences, and one with multiple focuses. They reported: "Our meta-analyses show that positive articles are cited about two times more often than negative ones. Our results suggest that citations are mostly based on the conclusion that authors draw rather than the underlying data."[9] Subsequently, Duyx et al. examined the literature on the "hygiene hypothesis"—a conjecture linking hygiene in early life to the subsequent development of allergies.[10] Studies that supported this hypothesis were cited about three times as often as unsupportive studies.

While most reports suggest a bias for statistically significant findings, Brian Hutchison found that the opposite was true of trials on the efficacy of pneumococcal vaccine.[11] Here, reviews of the literature were twice as likely to cite unsupportive trials as supportive trials. This suggests that it is not statistical significance per se that matters—but whether the findings conform to a favored claim.

One common recommendation is to use systematic reviews rather than the traditional narrative reviews.[12] Systematic reviews address a focused question in a way that attempts to make an exhaustive summary of current evidence combined with a critical appraisal of that evidence. They use a defined search strategy to capture not only all relevant published literature, but also, where possible, unpublished results. Systematic reviews of randomized controlled trials are now of key importance for evidence-based medicine. They use defined criteria to measure study quality, and the development of these criteria has led to progressive improvements in the quality and reliability of trials.

However, while systematic reviews make robust attempts to be unbiased, these are rare outside the sphere of clinical medicine. And, as Greenberg demonstrated (chapter 16), citing narrative reviews rather than primary studies can amplify biases.[13]

Citation Distortion

In 1980, Jane Porter and Hershel Jick published a five-sentence letter, "Addiction Rare in Patients Treated with Narcotics," in the *New England Journal of Medicine*. They had sifted through the records of 11,882 patients who had been prescribed at least one narcotic and had found only four cases of addiction:

> We conclude that despite widespread use of narcotic drugs in hospitals, the development of addiction is rare in medical patients with no history of addiction.[14]

This was potentially very important. As a letter, rather than a full peer-reviewed publication, it should have been regarded as an interesting observation rather than a definitive finding.

But in 2017, in another letter published in the *New England Journal of Medicine*, Pamela Leung and colleagues reported that Porter and Jick's letter had been cited in 608 papers between 1981 and 2017. They read each of these to see how it was cited; 439 of them (72 percent) had used it as evidence that, in patients treated with opioids, addiction is rare. Importantly, 491 of the citing papers failed to report that the letter had described the experience of patients that had been hospitalized; that is, patients in a well-controlled, safe setting under constant close supervision. Leung et al. concluded:

> A five-sentence letter published in the Journal in 1980 was heavily and uncritically cited as evidence that addiction was rare with long-term opioid therapy. We believe that this citation pattern contributed to the North American opioid crisis by helping to shape a narrative that allayed prescribers' concerns about the risk of addiction associated with long-term opioid therapy. Our findings highlight the potential consequences of inaccurate citation and underscore the need for diligence when citing previously published studies.[15]

In Greenberg's terminology, this was an example of "citation diversion," where the meaning of a study was subverted through chains of citation.

Sometimes references are simply copied from one paper to another. It is hard to know how common this is, but Pieter Kroonenberg, a Dutch statistician, discovered a nonexistent paper that had been cited more than four hundred times.[16] The phantom paper was cited as:

> Van der Geer, J., Hanraads, J. A. J., Lupton, R. A. 2010. The art of writing a scientific article. *J Sci. Commun.* 163 (2) 51–59.

This originated as a hypothetical example given in a style guide used by Elsevier to illustrate how to reference in particular journals. We confirmed, by a search in the Web of Science, that it had been cited more than 480 times by 2019. Most of these came from abstracts in conference proceedings, and it seems likely that many authors had misunderstood that this was an example of how to cite, not an example of something that should be cited. But the reference also appeared in seventy-nine journal papers. Of these, thirteen were connected together through references, and in these, it was bizarrely used to support the claim that a compound called rutin could dilute the blood, reduce capillary permeability and lower blood pressure. Here, it seems, this reference was simply being copied from one paper to another.

Underutilization of Evidence

To understand the extent to which the available evidence is utilized, Karen Robinson and Steven Goodman examined how often clinical trials were reported by the authors of later, similar trials.[17] They identified 1,523 trials and tracked how these had cited others on the same topic. Only about a quarter of the relevant trials were cited, which also constituted only about a quarter of the subjects enrolled in relevant trials. Strangely, a median of two trials were cited regardless of how many had been conducted; seemingly, while scientists may see further by standing on the shoulders of those who have gone before, two shoulders are enough, however many are available. The authors concluded, "Potential implications include ethically unjustifiable trials, wasted resources, incorrect conclusions, and unnecessary risks for trial participants."

Implications

The system of scientific communication appears to be more fragile than was once believed. It is impacted by researchers' decisions of what to publish and what to cite. Decisions made by individuals—individuals with particular research goals, expertise, finite memories, and an innate propensity for error—can affect what others believe in ways we still do not quite understand.

If evidence goes unpublished, it may rob others of the opportunity to critique a position. If citation bias is common, then scientists are likely to be basing their understanding on only a partial selection of the evidence. But while we may point to the waste in research time and funding if existing evidence is not fully utilized, and the inherent problems for the validity of scientific claims, the solution is not obvious. While these problems seem to be increasing, this might simply reflect greater recognition of them.[18] That greater recognition may itself drive self-correction in the behavior of scientists.

18 High-Impact Papers: Citation Rates, Citation Distortions, and Mis-citation

In 2017, *Nature* published a list of the hundred most cited papers of all time.[1] If how often a paper is cited tells of its importance, its quality, or its influence, then these, the top of the cream, should perhaps be allowed to speak.

The first thing to note is that thirty-two of these papers were published in journals with a 2017 impact factor (IF) of 3 or less, including the *Scandinavian Journal of Clinical and Laboratory Investigation* (IF 1.5) and the *Canadian Journal of Physics* (IF 0.9). *Nature* (IF 41.6) appears just four times—once more than *Science* (IF 37.2). The *Lancet* (IF 53.3) is there once, the *New England Journal of Medicine* (IF 79.3) not at all. The journal with the most entries is *The Journal of Biological Chemistry* (IF 4) with eight, followed by *Physical Review B* (IF 3.8) with seven. Fifty-seven different journals are represented on the list.

Nearly all of the papers are *methods* papers. In pole position is a 1951 paper from Oliver Lowry and colleagues that describes a simple way to measure the protein in a biological sample.[2] When a biochemist seeks to establish how much of any particular substance is present in a sample, they must relate it to the size of the sample. The crudest way is to weigh the sample, but measuring its protein content is more reliable. Lowry et al. has been cited more than 300,000 times, and another method of measuring protein content is number 3 on the list, with more than 150,000 citations.

Many methodological innovations are rapidly absorbed into conventional laboratory practice, but some have a more interesting trajectory. At number 61 on *Nature*'s list is "The Assessment and Analysis of Handedness: The Edinburgh Inventory."[3] Published in 1971, this paper was cited only fifty-eight times in the five years after publication, but nearly 1,500 times in 2018 alone. It's a scholarly paper which recognizes that any individual might use the right hand for some tasks and the left for others; it develops a ten-item questionnaire for measuring the degree of handedness, rather than treating handedness as a categorical distinction between left-handedness and right-handedness. But how has it come to be so widely used?

The answer is that this questionnaire is now used in brain imaging studies to make a categorical determination of whether the subjects are left-handed or right-handed.

It is probably not the case that, without this innovation, brain imaging studies would still be in their infancy, so perhaps we should be cautious about equating citation rates with influence.

"Thousands of Papers Cite Nonsense to Produce and Infer Nonsense"

Following the Matthew Effect, the rich become richer and a highly cited methods paper becomes ever more highly cited. Many researchers use a particular method because it is simple and because many others have used it in similar studies. But the physician-scientist John Ioannidis has argued that research tools that oversimplify complex problems can do harm. When the use of misleading methods is amplified to produce massive numbers of publications, "thousands of papers cite nonsense to produce and infer nonsense."[4]

In 2010, Andreas Stang published a commentary that criticized the Newcastle–Ottawa scale (NOS), a scale widely used in meta-analyses to appraise the quality of nonrandomized studies. The commentary pulled no punches, ending:

> In conclusions [*sic*], I believe that [the NOS scale] provide[s] a quality score that has unknown validity at best, or that includes quality items that are even invalid. The current version appears to be inacceptable for the quality ranking of both case-control studies and cohort studies in meta-analyses. The use of this score in evidence-based reviews and meta-analyses may produce highly arbitrary results.[5]

Then, in 2018, Stang and colleagues noted that, according to the Web of Science, Stang (2010) had been cited 1,250 times by the end of 2016.[6] They looked at exactly how it had been cited in a random sample of a hundred papers. Of these, ninety-six were systematic reviews and all but two of these wrongly portrayed the commentary as supporting the use of the scale in systematic reviews, exactly opposite to the unmistakable message of Stang (2010). It seems likely that a few meta-analyses started the quotation error and others followed by copying the citation without reading the paper.

Such citation copying may be common; one study developed a computational model to explain empirical studies of the prevalence of citations that repeat misprints of reference details, and from it they estimated that only about 20 percent of those who cite a paper have actually read it.[7] It is possible that the mis-citing authors had read the mis-cited paper, but merely copied the reference to it from another paper. However, the estimate is consistent with the findings of a meta-analysis of studies of quotation accuracy (that is, whether a reference is cited in a way that is consistent with its content), which estimated that about 12 percent of papers in medical journals contained major quotation errors—errors that either seriously misrepresented or which bore no resemblance to the cited source.[8]

By May 2019, Stang (2010) had been cited 3,271 times, according to Web of Science. The most highly cited paper that cites Stang (2010) is a 2013 meta-analysis that asked, "Are metabolically healthy overweight and obesity benign conditions?"[9] It cites Stang (2010) as follows:

> The Newcastle–Ottawa Scale (NOS) for assessing quality of nonrandomized studies in meta-analysis was applied [reference to the first description of the scale]. The NOS contains 8 items categorized into 3 domains (selection, comparability, and exposure). For each item, a series of response options is provided. A star system is used to enable semi-quantitative assessment of study quality, such that the highest-quality studies are awarded a maximum of 1 star per item; the exception is the comparability domain, which allows the assignment of 2 stars. As such, the NOS ranges from 0 to 9 stars [Stang (2010)].

This was typical of most citations to Stang (2010); it ignored the actual content. This paper concluded that "there is no healthy pattern of increased weight," and its findings contributed to a much larger meta-analysis from the Global BMI Mortality Collaboration (GBMC) that was published in the *Lancet* in 2018.[10] The GBMC study reported associations of both overweight (body mass index (BMI) 25–29.9) and obesity (BMI > 30) with higher all-cause mortality and claimed that these associations were broadly consistent in four continents. This, the consortium argued, "supports strategies to combat the entire spectrum of excess adiposity in many populations."

The GBMC study lists sixty-one authors in its writing committee and a massive number of participating researchers and consortia. Unusually, the study was accompanied by a commentary in the *Lancet* that warned about relying on observational evidence for policy recommendations.[11] It noted that few large trials had addressed whether weight-loss interventions reduce mortality or morbidity and noted that one trial had been ended after about ten years because it had found no association between weight loss and cardiovascular events.

Then in 2019, a paper by Katherine Flegal and a team ripped into the methodology of the GBMC study.[12] To reach its conclusion, they argued, the GBMC had used highly selected data, rather than a systematic approach. The GBMC had initially chosen individual participant data from 239 prospective studies with more than 10 million participants. It then excluded more than 60 percent of the data and more than 75 percent of fatal events by eliminating all cases with any reported disease at baseline or smoking history and all events within the first five years of follow-up. The effects of these restrictions was to reverse findings that overweight was associated with lower mortality and to strengthen the association of obesity with higher mortality. Flegal et al. concluded that "Given the major flaws in the selection process, in the adequacy of the data, in the data analysis, and in the interpretation, the GBMC conclusions should be viewed skeptically as a guide to action, either for clinical decisions or for public health in general. The flawed conclusion that overweight is uniformly associated with substantially increased risk of death and thus should be combated in any circumstances may lead not only to unjustified treatment efforts and potential harm in a wide range of clinical conditions but also to a tremendous waste of resources."

Flegal and colleagues noted also that the GBMC paper had claimed, incorrectly, that the findings were broadly consistent with the analyses done in a 2015 study of 12.8 million Korean adults by Sang-Wook Yi and colleagues—a study of a sample larger than all 239 GBMC datasets combined, but which was not included in the GBMC meta-analysis.[13] Yi et al. had reported that overweight was not associated with increased mortality and that moderate obesity (BMI 30–35) was associated with only a very small increase. Even after applying similar restrictions to those used by the GBMC, Yi et al. still found no increased mortality in overweight subjects. The conclusions were not consistent with simple public health advice. Women, especially younger women, had a lower optimal BMI than men, the optimal BMI increased with age, and optimum BMI for both sexes and all ages except younger women was higher than the current definition of "normal weight" (BMI 18.5–24.9).

The implicit allegation by Flegal et al. is that the GBMC had selectively chosen studies to reach a preformed conclusion in support of the current definition of obesity by BMI, a definition introduced more than thirty years ago, and that they mis-cited discrepant evidence to strengthen that conclusion.[14]

The BMI is a simple measure, and the cut-offs for defining overweight and obesity are memorably simple—all positive features for using it in public health advice. On January 5, 2013, the *Economist* published a letter from Nick Trefethen, professor of numerical analysis at the University of Oxford. He noted that people don't scale in a linear fashion as they grow:

> SIR—The body-mass index that you (and the National Health Service) count on to assess obesity is a bizarre measure. We live in a three-dimensional world, yet the BMI is defined as weight divided by height squared. It was invented in the 1840s, before calculators, when a formula had to be very simple to be usable. As a consequence of this ill-founded definition, millions of short people think they

are thinner than they are, and millions of tall people think they are fatter.[15]

Here we are not adjudicating between the GBMC study and Flegal et al. We are witnesses, not judges, and have noted that Flegal et al. had a prior investment in their own hypothesis, but we have not interrogated this. Our point is that we cannot judge the quality of a paper by how often it has been cited, nor by the journal in which it is published, nor by the number and authority of the authors. There is no short cut. The strength of the evidence within any paper has to be judged by the reader, and it is a judgement that does not serve forever but is contingent on whatever new evidence may emerge. How well the strength of evidence is weighed will depend in part on how well the reader perceives the authors' biases—and no less, in how well the reader knows their own biases and compensates for them. That is the challenge of scholarship.

19   Are Most Published Research Findings False? Weaknesses in the Design of Experiments and the Analysis of Their Outcomes

In 2005, John Ioannidis published an essay with the startling title "Why Most Published Research Findings Are False."[1] He argued that claims are more likely to be false when the studies are small; when the effect sizes are small; when more relationships are tested; when fewer of the tested relationships are preselected; when there is more flexibility in designs, definitions, outcomes, and analytical modes; and when financial or other interests are involved.

Claims Are More Likely to Be False When the Studies Are Small

There are many reasons why small studies can be unreliable. But think of this "experiment" in psychokinesis: think of an animal—a rabbit—and toss a coin three times. Say it comes up HTH. Now think of another animal—a squirrel—it comes up TTT. Three heads or three tails will come up quite often in a sequence of three tosses—one in four times. But when TTT coincides with a squirrel, you formulate a hypothesis—thinking of a squirrel's bushy tail makes it likely that the coin will come up tails whereas thinking of a rabbit does not. So think of a squirrel again and toss the coin—there's a 50/50 chance of a T, and if it does you have TTTT and you begin to believe your hypothesis—just one more throw—again a 50/50 chance and it's TTTTT. There is just a 1 in 32 chance of getting TTTTT by chance alone—well within your threshold of $P < 0.05$ for a significant result. Now think of a rabbit and toss the coin a couple more times—you won't get a significant result as you've already got HTH. So you end up with a significant effect of thinking about a squirrel and no significant effect of thinking about a rabbit.

You are now invested in the hypothesis—you've convinced yourself. You might be tempted to try one more flip of the coin. With again a 50/50 chance you get TTTTTT—now a result with just a 1 in 64 probability of getting by chance. Maybe the sixth throw is a H—an outlier—but wait, weren't you distracted on that sixth throw, didn't the phone ring? Discard it.

Let's break down what went wrong:

1. The hypothesis was not prespecified—it emerged only from the outcome of the pilot experiment.[2]

2. The results of the pilot were treated as part of the definitive test, biasing the outcome of the statistical test. Few papers report pilot studies separately—they either do not do pilot experiments, which seems unlikely, or they do them but don't report them, or they absorb the results of pilot experiments into the reported outcomes as though they were part of a definitive hypothesis-testing experiment.

3. The study was stopped when it just reached statistical significance—you didn't risk upsetting a nice outcome with any more throws. This is thought to be common as inferred from an excess in the literature of studies that just reach the $P < 0.05$ threshold.[3]

4. The effect was declared to be selective because thinking of a squirrel produced a significant effect but thinking of a

rabbit did not. This is a common flaw in reasoning in scientific papers—a difference between two treatments is declared without directly comparing the treatments.[4]

5. An outlier was discarded for no good reason other than it didn't fit. There can be good reasons to discard outliers, but it's important to mention them and explain them. It's seldom done. Often the only way of detecting that outliers have been discarded is because of an unexplained disparity in the reported n values. In a survey of 3,247 scientists, 15 percent admitted to "dropping observations or data points from analyses based on a gut feeling that they were inaccurate" in the previous three years.[5]

Finally, retrospectively construct a rationale for the experiment. Trawl the literature to find previous positive experiments on psychokinesis. Note the salience of a squirrel's tail—its prevalence in children's literature, its emotional and cultural connotations. Note previous studies of emotional salience and its effects on brain activity: mention a few brain sites (the amygdala is a good one) and mention fMRI studies that show its involvement in emotion, intentionality, and motor control. Construct a plausible sounding hypothesis and an evocative title—"Will and the Salience of Mental Imagery." Publish.

Claims Are More Likely to Be False When Effect Sizes Are Small

"Effect size" simply means the difference between a control measurement and the measurement after an intervention. But in this context, "effect size" has a particular technical meaning: it refers to "standardized" effect size, where the effect size is scaled relative to the variance of the measurements. Thus "small" effect size is not a statement about the absolute size of an effect, but a statement about its size relative to the variation expected in comparisons where there is no true effect.

In small studies that report an unexpected outcome, only when the effect size is large is it likely to be a true effect—but the effect size reported in small studies is likely to exaggerate the true effect size. This arises from the so-called winner's curse. Even where there is a true effect, a small study is not likely to find it to be statistically significant unless, by chance, the data show a larger effect than the true effect.[6]

It's not enough for a study to be big enough for a "just significant effect"—it needs to be reliable and replicable. This requires doing a *power analysis* before the experiment. By this, scientists calculate how large the sample must be to have a good chance of detecting an effect of a specified size at a given level of significance. If an experiment is adequately powered, its outcomes are likely to be replicable. But very few are adequately powered.[7]

Claims Are More Likely to Be False When More Relationships Are Tested

This is the issue of "multiple comparisons" mentioned elsewhere. These problems arise in many areas of science and might be suspected when surprising conclusions appear in a study involving many variables. For example, one study, "You are what your mother eats," attracted considerable media attention because of its widely reported conclusion that having cereal for breakfast enhances a woman's chance of conceiving a boy.[8] The authors' own conclusions were more considered; they interpreted the outcome as indicating that energy deficiency in the mother at the time of conception was a signal of harsh environment and that this led to a biologically determined selection for male offspring. The paper provoked a prompt attack on the basis that so many comparisons were implicit in the design of the study that while the reported P value for breakfast cereal was an impressive 0.003, adjusting for multiple comparisons would reduce this to just 0.28.[9]

The notion that an energy-deficient environment would favor the production of male children was subsequently examined by studying the birth records of children conceived in the Netherlands during the Dutch famine of 1944–1945. It showed no association between energy intake and sex ratio.[10]

Claims Are More Likely to Be False When Fewer of the Tested Relationships Are Preselected

Scientists often do not pursue just one preselected hypothesis, but take an expansive approach, trying many things, collecting diverse data, and hunting for relationships between them. By this they are implicitly exploring many hypotheses, mostly unstated or unconceived before the results are in. The chance of getting a false-positive outcome in some of these by misapplying hypothesis-testing statistics is high.

Claims Are More Likely to Be False When There Is More Flexibility in Designs, Definitions, Outcomes, and Analytical Modes

This covers such a vast range of "questionable practices" that it is hard to know where to start. Broadly, this concerns how, at the end of a complex study, the data are interrogated to find some way of presenting outcomes as significant. To this end, some sets of data may be excluded, some may be split into subsets and others combined, and criteria for deciding whether an effect has occurred may be redefined. Different types of tests may then be tried to find one that produces an acceptable P value, with minimal consideration of the assumptions that underlie its validity.

In 2011, noting that many papers had documented how adept people are at interpreting ambiguous information to justify conclusions that mesh with their desires, Joseph Simmons and colleagues reflected on how papers in psychology had used different criteria to classify their observations. They showed, by computer simulations and in experiments, "how unacceptably easy it is to accumulate (and report) statistically significant evidence for a false hypothesis."[11] The remedy is not necessarily to abandon statistical significance but to insist on the necessity of defining, before the data have been collected, exactly how those data will be analyzed.[12]

Claims Are More Likely to Be False When Financial or Other Interests Are Involved

Published studies often (but not always) favor outcomes that are preferred by the funders of the study, and this has been well documented in the case of drug trials funded by pharmaceutical companies.[13] This does not arise because such trials are of low quality or fraudulent, but because many trials are designed in a way that favors a supportive outcome. This can be done, for example, by carefully choosing how to measure efficacy, by comparing the trial drug not with the best alternative but with an inferior competitor, by choosing the doses to compare, by very large trials that can detect trivially small differences, by selecting patients most likely to respond to treatment, and by selectively publishing positive studies.

For example, use of such tactics in studies appeared to show, in industry-sponsored head-to-head comparisons of three anti-psychotics, that one drug (olanzapine) was better than another (risperidone), which was better than another (quetiapine), which in turn was better than olanzapine.[14]

Such conflicts can also affect supposedly unbiased systematic reviews. One study looked at reviews on sugar-sweetened beverages and weight gain: systematic reviews written by authors with food industry funding were five times more likely to conclude that there is no positive association than reviews by authors with no such funding.[15]

It is easy to see possible bias in papers where the authors declare a conflict of interest. However, as we have argued, all authors have a conflict of interest. All are proponents of their own interpretation of evidence and all have invested in certain claims. This potential for bias in any paper and any review is inescapable. The only defense is transparent acknowledgement of potential bias, for authors to be aware of it and to compensate for it by attempts at objectivity, and for readers to be critically alert.

"Statistics" merely refers to the quantitative description of data. Formal statistical "testing" was introduced by Ronald Fisher. In his classic work, *The Design of Experiments*, Fisher noted that, when any scientific claim is attacked, its critics argue either that the outcomes of experiments have been misinterpreted, or that the design of the experiments was faulty. These come down to the same thing—if the design of an experiment is faulty, any way of interpreting its outcome is faulty. Experiments obviously must be designed before they are performed, and good design involves considering what outcomes are possible and how to distinguish a "true effect" from a chance difference. For this, the professional statistician has a necessary role, but Fisher insisted that "the questions involved can be dissociated from all that is technical in the statistician's craft, and when so detached, are questions only of the right use of human reasoning powers. [. . .] The statistician cannot excuse himself from the duty of getting his head clear on the principles of scientific inference, but equally no other thinking man can avoid a like obligation."[16]

In Fisher's day, and for long after, few technical approaches were available to scientists, and few means of gathering and analyzing large amounts of data. Experiments were time consuming and had to be designed carefully. Scientists often built their own equipment for a particular purpose or designed it for building in departmental workshops. By this, they understood their apparatus in intimate detail. They demonstrated their apparatus to others, inviting comments and criticism. Their experiments were designed to be simple and transparent, and the effects that they reported had to be large and readily reproducible to be convincing. The older literature includes a rich archive of robust and reliable observation and examples of thoughtful experimental design.

By contrast, today's scientists often do not understand well the equipment and methods that they use; equipment is bought and methods copied from other papers. Experiments are designed to meet the competence of the equipment, rather than equipment being designed to serve the purposes of the experiment. Sometimes equipment or methods were designed for one purpose but co-opted for another to which they are ill suited. Such is the diversity of methods and equipment that is used and the diversity of purposes to which they are applied, and so common is the ignorance of them, that scrutiny of methodology is inconsistent.

This is also true of statistical methods. Few journals use professional statistical editors, and statistical flaws are abundant in the literature. Such is the strength of concern that some journals have banned hypothesis-testing statistics and P values from their pages, requiring that authors confine themselves to reporting effect sizes and confidence limits.[17] Others favor abandoning assertions about statistical significance on different grounds—that it is invalid to conclude there is "no difference" or "no association" just because a P value is larger than 0.05 or because a confidence interval includes zero. However, some concept of significance, not necessarily statistical significance only, seems to be essential for science to avoid being overwhelmed by noise in the literature.[18]

Scientists are attracted to hypothesis-testing statistics in part from a misapprehension that they are pursuing "hypothesis-driven science" and are following the spirit of Karl Popper. Nothing could be further from the truth. For Popper, Science should be driven by hypotheses; he did not suppose that experiments could be driven by the need to construct hypotheses. Popper argued that Science was about testing bold hypotheses, by attempting to refute them: hypothesis-testing statistics can refute only the null hypothesis, and a hypothesis that nothing will happen is seldom bold.

In the 1970s a paper in the *Journal of Physiology* might have displayed one or two P values, but often none. Today it is not uncommon to find fifty or more P values, sometimes with no descriptive statistics at all associated with them. It is true that to give an effect size with a confidence level is sometimes equivalent to stating a P value: where that is true, it seems good reason to omit the *P* value. If an effect is clearly present, to assert that it is a significant effect is not merely superfluous, it is a rhetorical flourish that exaggerates its importance. When the data are given prominence, it forces attention on how big the difference is, and whether it is significant in a more meaningful sense.

Of course, there is an important place for hypothesis-testing statistics—for testing transparently preconceived hypotheses with experiments transparently designed to do exactly that, using tests transparently designed in advance.

As we have argued, much older science was exemplary, and much of it was hypothesis-driven in the sense that Popper would have recognized. But early clinical and biomedical studies, which trialed interventions with no mechanistic understanding that might inform a sound design, were not. The quality of medical science has been transformed by awareness of principles of good trial design that yield plausible grounds for inferring causality even without strong mechanistic understanding. It is now accepted that definitive clinical trials should be large, multicenter trials, objectively randomized, preregistered with defined outcome measures, properly controlled, with transparently declared primary outcomes and a prespecified strategy for statistical analysis, and properly blinded throughout. Such large trials are expensive and have to be reserved for questions of particular importance where the evidence is uncertain. But the principles that inform such gold-standard design have infiltrated the conduct of many studies.

However, these criteria are not simply translatable to all areas of research; each area needs its own relevant criteria of quality. Good experimental design is not a given set of rules. It is a discipline of reasoning.

20 Societal and Economic Impact of Basic Research

> There is an absurd notion about that education and health are things that we spend our national wealth on. What an extraordinary idea this is that money has any value in itself; education and health are not things we spend our wealth on, they are our wealth. To be a scientist, to be a creator of knowledge and understanding, is to be a creator of wealth in the most important and enduring sense.[1]
>
> —Gareth Leng, *The Heart of the Brain*

We began chapter 1 by saying "If you ask any scientist what it is that scientists do, she might answer that scientists are engaged in extending our knowledge and understanding of the world." We now might begin to see what is so odd about this answer. A chef cooks meals, a carpenter makes furniture, a doctor diagnoses illness and prescribes medicine, a lawyer pleads cases in court—they will all probably describe what they do in terms of observable outputs, outputs that affect other people in specific ways. What any scientist does, in similar terms, is to write papers for scientific journals and persuade other scientists of their importance. This might extend our understanding and knowledge much as meals might (or might not) fill us with contentment, furniture might beautify our homes, and medicines might cure our ailments. But the chef, the carpenter, and the doctor have many witnesses to the impact of their outputs; the scientist has, usually, only other scientists. We can measure the output of scientists only by the papers themselves. We might try to measure the economic and societal impact of those papers—and such attempts are increasingly common—but these attempts involve, as we shall see, constructing another, highly contestable, narrative.

If we reflect on the history of oxytocin as told in chapter 13, we might be forgiven for not noticing what oxytocin is most important for. That question has a clear answer if we understand it as asking what oxytocin is useful for.

Oxytocin has been, for at least seventy years, an indispensable part of the normal management of childbirth. In developed countries, most women giving birth are given oxytocin either to facilitate childbirth, to aid the expulsion of the afterbirth, or to prevent postpartum hemorrhage. There has been no pause in the work of scientists to refine this application; in 2018 alone, nineteen papers reported the results of clinical trials of oxytocin in labor wards, reflecting on the indications and contraindications for its use and comparing it with alternative medications.

One of those papers, in the *New England Journal of Medicine*, reported in 2018 on a large, multicenter, randomized clinical trial conducted by the World Health Organization; it involved nearly 30,000 women in twenty-three countries.[2] It looked at postpartum hemorrhage, a serious condition that is mainly due to poor contraction of the uterus after childbirth and which can be prevented by drugs that enhance uterine contractions. In the last fifty years, large reductions in maternal mortality have been achieved in most countries, mainly through the use of oxytocin, the current standard therapy. Nevertheless, between 2003 and 2009, postpartum hemorrhage accounted for more than 400,000 deaths worldwide, mainly in developing countries. Oxytocin, to be fully effective, must be stored between 2°C and 8°C, and in many countries this is a considerable barrier to its use.

The trial compared oxytocin with a long-acting oxytocin agonist, carbetocin, and showed that carbetocin and oxytocin were equally effective in preventing postpartum hemorrhage. Unlike oxytocin, however, carbetocin can be produced in a way that makes it heat-stable. Avoiding the need to keep it cold means lower transport and storage costs and less waste, and it can be used in many more settings worldwide.

Carbetocin is not new; its use was first described in pigs in 1981. PubMed catalogues 189 papers on carbetocin, papers that have, over forty years, documented its chemical properties, its biological activities, and its safety.

This path, from an observation in basic science to a treatment that can prevent the deaths of thousands of women in childbirth, exemplifies the long and uncertain trajectory of translational impact. Those 189 papers were invisible to the search that we described in chapter 13; they were not highly cited, but known only to a few who pursued a distant and uncertain promise.

We cannot know, of the papers published in the last few years, which, if any, will have similar impact. We cannot know which, if any, are important, in the sense that most of us will recognize that it is important to prevent deaths of women in childbirth. Will oxytocin offer new treatments for pain, for weight loss, for autism, for osteoporosis, for cardiac health—or for none of these or for something else entirely? Nobody sensible can say for sure.

But while nobody sensible can say that they will lead to new therapies, increasingly papers are advertising the possibility that they might.

Scientists now in all domains of research are under pressure to maximize "societal and economic impact" from their research, and on the face of it there seems little reason to argue with this. When the public pays for the research, it would seem only reasonable that they should be told what scientists do with their involuntary investment, what the returns are, and whether they might be greater.

This is far from easy to deliver. The socioeconomic benefits of science arise in many different domains, most of which are very hard to quantify at all and often impossible to quantify in any way that any sensible person would regard as reflecting their value. Nor can we expect any two people, chosen at random, to agree on that value. How do we calculate and weigh the value of saving a mother from dying in childbirth, of reducing our carbon footprint by creating better ovens to bake bread, of inspiring wonder and increasing understanding of the world and worlds beyond? The problem with metrics is that we measure what we can measure and ignore what we can't. We wouldn't bother with metrics if not for some purpose, and that purpose is usually to decide what to keep and what to throw away. But perhaps we should decide on purpose first and choose the metric to suit it.

Whenever a metric is introduced and understood to place some value on the activities of people, one thing is likely to ensue: the people concerned will change their behavior. Of course, that is generally the point—by measuring research success to improve research success. But when the metric measures some surrogate of that success, the behavior change is directed towards the surrogate, not the thing that it's intended to measure.

Impact Statements

In the United Kingdom, funding for research is allocated to universities partly on the basis of impact statements that make claims for economic benefits that have arisen from their past research activity. In addition, many research funders now require that applications for funding be aligned to priority areas, topics perceived to be in particular need of more research. They require that applications be accompanied by an "impact statement" that contains promises of benefit, and a strategy for achieving it.

It is not only funders that are increasingly aware of impact in this sense. Journals promote selected papers for their claims for socioeconomic impact. Academic institutions promote these claims by press releases. Scientists have always been willing to take any opportunity to talk to an interested public, and, as science festivals thrive, opportunities can be abundant—especially for those with an exciting message of hope. That scientists engage with the public is required by some funders and is an expected part of the impact plans that they must construct.

This is a truth-toxic environment. If scientists peddle false hopes, they will undermine respect for science; if they make inflated claims for importance, the credibility of all science will suffer. If they want the words that they write and speak to be taken seriously, it is dangerous to decorate them with hype. Yet such hype is exploited by journals, celebrated by institutions and funders, and propagated by the media.

The reality is complex. Oxytocin passed swiftly into clinical application because its use could be confined to circumstances in which basic science could establish that it was physiologically relevant. It could be used at low doses because basic science had established that at the relevant times, the uterus was

exceptionally sensitive to oxytocin, and these low doses mini-
mized the possibility of "off-target" side effects. Complications
of actions within the brain did not arise, because at these doses,
as basic science had established, the brain is protected from
injected oxytocin by a blood–brain barrier.

But the central actions of oxytocin, its actions on the brain,
offer hope for wholly different therapeutic applications—
applications that were inconceivable before basic science had
extended our understanding. In at least three different areas,
oxytocin offers such hope, in each case for major health conditions that have been intractable to present therapies. First,
the involvement of oxytocin in appetite regulation and glucose
homeostasis offers opportunities for interventions in weight
management and diabetes. Second, its involvement in social
behaviors offers opportunities for therapies for diverse conditions of disordered social behavior, including autism spectrum
disorder. Third, its role in pain communication offers hope for
new treatments for intractable pain—there is some evidence that
it might lead to treatments for migraines. Nor are these the only
possibilities; its role in bone remodeling and in cardiac development raise prospects for therapeutic exploitation in these areas
also, as does its potential application to conditions of sexual dysfunction. But these are hopes, and hopes are often dupes. For
every drug that reaches the clinic as an effective new treatment,
many, many more will fall by the wayside.

The difficulties involved in translation are daunting. As oxytocin affects many systems, how can it lead to an effective treatment for one particular condition without unintended and often
unwanted side effects? From a "lead compound," chemists can
do clever things; they can make it more stable, more selective,
more "bioavailable" so that it can pass membranes more readily

and survive degradation in the gut for example. Basic neuroscientists can now conceive of ways in which drugs might be delivered to specific targets in the brain—and biologists can conceive of work-around solutions, by targeting not oxytocin itself but the cells that produce it, to stimulate them to secrete more, or by targeting the cellular pathways by which oxytocin acts. Every one of these possibilities is the subject of basic research, perhaps only one will prove feasible for translation to the clinic, perhaps none will. If we knew which one could work, then by harnessing the collective efforts of the global multidisciplinary community of researchers we might accelerate translation radically. But we do not know which, if any, will work. So we must keep our options open, pushing at the edges of our knowledge and understanding across a broad front, looking for windows of opportunity to open, hoping, waiting all the time for new unthought-of opportunities to arise from other, apparently unrelated areas of basic science. This is what basic scientists do, and they do it because this is what must be done. In Arthur Clough's lines again,

> If hopes were dupes, fears may be liars;
> It may be, in yon mist concealed,
> Your comrades chase e'en now the fliers
> And, but for you, possess the field.[3]

It is not only by finding a way forward that we make progress: finding the blind alleys and closing them down is no less important, especially for those who may follow, perhaps years later. Pharmaceutical companies know this well: it is a misconception that, when they engage basic scientists in collaborations, they are only interested in supportive results. More important to them is to know as soon as possible when a lead is destined to fail.

As should be self-evident, the link between basic science and actual benefit is unpredictable and often far distant yet ultimately, at some level, all benefits depend on advances in basic science.

The case for this was made by Abraham Flexner (1866–1959) in an essay in 1939, "The usefulness of useless knowledge." Flexner was an educator, not a physician, but is best known for his role in reforming medical education in the United States and Canada. He began his essay by recounting a conversation with the philanthropist George Eastman, who was then contemplating devoting his vast wealth to the promotion of education in "useful subjects." When Flexner asked him who he considered to be the most useful scientist in the world, he replied, "Marconi." Flexner's response was:

> Mr. Eastman, Marconi was inevitable. The real credit for everything that has been done in the field of wireless belongs, as far as such fundamental credit can be definitely assigned to anyone, to Professor Clerk Maxwell, who in 1865 carried out certain abstruse and remote calculations in the field of magnetism and electricity. [. . .] Finally in 1887 and 1888 the scientific problem still remaining—the detection and demonstration of the electromagnetic waves which are the carriers of wireless signals—was solved by Heinrich Hertz [. . .] Neither Maxwell nor Hertz had any concern about the utility of their work; no such thought ever entered their minds. They had no practical objective. The inventor in the legal sense was of course Marconi, but what did Marconi invent? Merely the last technical detail, mainly the now obsolete receiving device called coherer, almost universally discarded.[4]

The Societal Impact of Research

In 1910, Flexner had published a caustic critique of the state of higher education in the United States.[5] He attacked the role of lectures in university teaching, which enabled colleges to "handle cheaply by wholesale a large body of students that would be otherwise unmanageable and thus give the lecturer time for research." He argued that "research had largely appropriated the resources of the college, substituting the methods and interest of highly specialized investigation for the larger objects of college teaching." His book led the Carnegie Foundation to fund him to study medical education in Canada and the United States, and two years later he published what came to be known as the Flexner Report.[6]

Flexner had visited all of the medical schools in the United States and examined their entrance requirements, the size and training of the faculty, their financial resources and income, the quality of their laboratories, and their access to a teaching hospital. He concluded that, while the schools aspired to scientific principles of medical education, few had the means to deliver them. An abundance of scarcely regulated commercial schools massively overproduced physicians of very variable, but generally poor quality.

Flexner argued that the commercial ethic that then governed "proprietary" medical schools was not compatible with the academic values necessary for medical education: "The overwhelming importance of preventive medicine, sanitation, and public health indicates that in modern life the medical profession is an organ differentiated by society for its highest purposes, not a business to be exploited."[7]

Flexner proposed the "speedy demise" of most of the schools and made far-reaching recommendations for curriculum reform. He argued that medical education could not be a matter left to the rote learning of facts, but had to be an exercise in developing those critical and analytical skills that characterized the research scientist:

> The scientist is confronted by a definite situation; he observes it for the purpose of taking in all the facts. These suggest to him a line of

action. He constructs a hypothesis, as we say. Upon this he acts, and
the practical outcome of his procedure refutes, confirms, or modi-
fies his theory. Between theory and fact his mind flies like a shuttle;
and theory is helpful and important just to the degree in which it
enables him to understand, relate, and control phenomena. [. . .]
The physician, too, is confronted by a definite situation. [. . .] The
patient's history, conditions, symptoms, form his data. Therefore he,
too, frames his working hypothesis, now called a diagnosis. It suggests a line of action. [. . .] The sick man's progress is nature's com-
ment and criticism. [. . .] The progress of science and the scientific or
intelligent practice of medicine employ, therefore, exactly the same
technique.[8]

The report transformed the nature of US medical education
by consolidating it in a much smaller number of well-funded
schools, and by embracing "scientific knowledge and its
advancement as the defining ethos of a modern physician."

In the years that followed, the report was blamed for fostering a crowded, inflexible medical curriculum.[9] This is not what
Flexner sought: he had explicitly railed against the "absurd over-
crowding" of the medical curriculum and warned of the dangers
of rigidity. Nor did he ignore the doctor–patient relationship
and the humane aspects of medical care. In 1925, frustrated at
its neglect in the new curricula, he wrote, "Scientific medicine
in America—young, vigorous, and positivistic—is today sadly
deficient in cultural and philosophical background."[10] Nor did
he ignore preventative medicine: he declared that doctors must
remember that disease commonly is the consequence of unfavor-
able environmental conditions and that doctors have a duty "to
promote social conditions that conduce to physical well-being."

The recognition that higher education is not about delivering
facts but about delivering skills of thought and about placing
knowledge in a broad context against a background of social,

cultural, and philosophical awareness, is a maxim to most of those involved it its delivery. If its value is neglected, that is a reflection of the inadequacy of metrics in many domains—not least because it is easy to measure knowledge in depth but harder to measure breadth of knowledge and harder still to measure the quality of thought. Nevertheless, if research in any university does not make a critically important contribution to the quality of the education that its students receive, then there is surely something very wrong either with the way that research is conducted in that university or in the way that its teaching is delivered.

Research in its different forms can and does have societal impact in many domains, in its influence on policies, on public awareness and understanding of issues that concern them, on the structures and functions of its various institutions in all sectors of life. But this most direct, most universal, arguably most important societal impact, on higher education itself as well as the critical analytical skills of its graduates, is often forgotten.

Economic Impacts

For some areas of science, economic returns can be calculated. For applied sciences, universities can point to patents, spin-off companies, and partnerships with industry. For basic sciences, it is harder, because of the long lags between research and translational returns and because of the tortuous and contestable paths that connect basic science to specific innovations.

Total economic returns are complicated to calculate. The direct returns from medical research comprise the health gains minus the health care costs of delivering them. This requires putting a monetary value on the health gains. A second, indirect,

return comprises gains in national income that result from the
research and from the activity that it stimulates. In the case of
medical research there is a strong symbiotic relationship with
pharmaceutical industry. The academic literature on the medical and biotechnology sectors shows that public research and
private research and development are complements, not substi-
tutes; public research stimulates private research, and vice versa.
Both lead to improved productivity in the economy generally.

In the United Kingdom in 2006, the Academy of Medical Sciences, the MRC, and the Wellcome Trust commissioned research
to compare the benefits accruing from UK medical research with
its costs.[11] Previous work, mainly in the United States—the so-
called exceptional returns literature—had highlighted economic
and health gains associated with public investment in basic
research.[12] Could this be extrapolated to the United Kingdom,
and how could this be tested rigorously and transparently?

Several studies had converged on the estimate that it takes an
average of seventeen years for research evidence to reach clinical practice, so this research took a long view. It estimated the
economic benefit of research on cardiovascular disease between
1975 and 2005. Using an estimated lag of seventeen years, it
concluded that every £100 invested in public/charitable research
on cardiovascular disease produced benefits equivalent to £39
per year in perpetuity. Of this, £9 was attributable to the benefit from health improvements, and the remaining £30 to "spillovers" benefiting the wider economy.

The study identified clinical interventions that had arisen
from medical research whose benefits had been evaluated.
Clinical benefits are commonly measured in terms of "quality-
adjusted life years" (QALYs), where one QALY is equivalent
to one year of life in perfect health. Benefits are calculated by

estimating the years of life remaining for a patient following a particular intervention and weighting each year with a quality-of-life score that is measured in terms of the person's ability to carry out normal activities and freedom from pain and mental disturbance. QALYs were attributed a monetary value of £25,000, using a formula generally applied by the National Institute for Health and Clinical Excellence. The study estimated the total value of the QALYs gained from specific interventions between 1985 and 2005; it came to a best estimate of £69 billion (at 2005 prices), with upper and lower estimates of £91 billion and £55 billion. The health care costs associated with those gains over the same period amounted to £16 billion (£11–17 billion). Thus, the net direct benefit was about £53 billion.[13]

The report estimated the expenditure on cardiovascular research from public and charitable funders by analyzing grants awarded. In 2005 prices, annual funding for cardiovascular research fell from £144 million in 1975 to £121 million in 1992, with a total expenditure over the period of £2 billion. Thus the research costs of £2 billion could be associated with a net benefit of £53 billion.

These benefits cannot be attributed wholly to UK research—medical research is a global enterprise. The proportionate contribution from the United Kingdom was estimated from an analysis of citations in UK clinical guidelines in cardiovascular disease, combined with the findings of previous studies. This led to the estimate that the proportion of UK health care benefit that could be attributed to UK research on cardiovascular disease was somewhere between 10 percent and 25 percent, with a central estimate of 17 percent equivalent to about £9 billion.

However, not all UK research was funded by public and charitable causes. In this same period, pharmaceutical industry spending on cardiovascular research in the United Kingdom greatly exceeded public and charitable funding—by 2.4 times at the midpoint of 1992. Taking this into account yields the conclusion that the public investment in cardiovascular research, after a lag of seventeen years, yields a 9 percent return on investment in perpetuity.

But this return, healthy though it seems, is still not the major return. "Spillovers" comprise the return on an investment to other organizations in the same sector and the return to all other parts of the economy. These were calculated (1) by analyzing the relationship between public/charitable and private R&D and then the relationship between private R&D and gross domestic product, and (2) using the economic literature to estimate the social rate of return to public R&D. This led to the estimate that the economic gains from public/charitable medical research deliver an additional rate of return of 30 percent. This estimate has a wide margin of uncertainty (20–67 percent), and as the evidence behind it came mainly from studies in the United States, only some of which was specific to medical research, its application to the United Kingdom and to medical research is tentative, but this is the best evidence available, and subsequent studies have come to similar conclusions.[14]

Given that applied research rests upon basic research and that even the narrow economic returns of research are considerable, the question might be asked what role should the public purse have in this enterprise—should it not be left to the private market to make that investment and enjoy its returns? The simple answer is that, for basic research, it is inherently unpredictable where those returns will arise; they cannot be corralled by private investors. In the United Kingdom, the government of Margaret Thatcher in the 1980s saw this clearly. This was a government that took a scythe to public spending, and research did not evade its bite. In those years, much research was conducted in government-funded research institutes, and the Thatcher government undertook to identify which research conducted in those institutes was "applied research," with an intended economic benefit, and which was fundamental basic research. The exercise sent a shiver of apprehension down the backs of basic scientists, but that apprehension was partly misguided. For, with the results of the survey established, the government then undertook to withdraw from all near-market research, on the basis that the private sector was better placed to judge the value of investing in it, while recognizing the duty of a responsible government to safeguard the fundamental research on which future prosperity would depend.[15] This did not, however, mean that basic research funding was maintained; in computer science, physics, and biology, funding continued to decline.[16]

Applied research fell between the cracks, and here at least, some consequences of the Thatcher scythe proved to be unfortunate. One area of applied research that was eliminated was research into scrapie, a peculiar disease found in sheep that had no human counterpart and little prospective benefit except to sheep farmers, a small community, disorganized, and economically negligible at a time when environmental, social, and cultural benefits were disregarded. Within a few years, bovine spongiform encephalopathy—mad cow disease—hit UK agriculture like a storm. The infectious agent proved to be closely associated with the agent of scrapie, but the United Kingdom's expertise in scrapie, which comprised most of the global expertise, had been eliminated. The disease began to spread to humans, and in a panic, the stock of cattle in the United Kingdom went in vast bonfires.

Citation Networks

This type of analysis can give a broad appreciation of the impact of basic science. But citation networks provide a way to trace the trails from particular pieces of basic science to particular outcomes. For example, one study sought to identify the trail of research that had led to regulatory approval for two new drugs: ipilimumab in oncology and ivacaftor for cystic fibrosis.[17] Ipilimumab induces sustained remissions in patients with previously intractable cancers by activating immune effector cells; ivacaftor corrects a mutation in the product of a gene that is affected in cystic fibrosis, and it is the first targeted therapy for this disease. Beginning with the references cited in clinical trials and in information provided to the regulatory authorities, the authors reconstructed the networks of articles, authors, and institutions that had contributed to the success of these two drugs. For ipilimumab, the process identified 7,067 scientists with 5,666 different institutional and departmental affiliations and spanned more than a hundred years of research. For ivacaftor, it identified 2,857 scientists from 2,516 addresses, spanning fifty-nine years of research.

For the many reasons that we have discussed, these citations can neither exhaustively capture the contributions of scientists to these discoveries, nor will all that they do capture have played equally important roles. But such networks can be analyzed to identify hubs that have contributed either to key discoveries, or to conceptual insights, or to effective synthesis and intercommunication. For ipilimumab, fifteen scientists and seven institutions associated with 433 articles over forty-six years served as hubs for 31 percent of the network. For ivacaftor, thirty-three scientists and seven institutions associated with 355 articles over forty-seven years served as hubs for half of the network.

These data quantify how the knowledge base on which important advances in medicine ("cures") depend includes contributions from a large and diverse set of individual scientists working in many locales. This insight should be instructive for policy makers by suggesting that future cures will depend on broadly based public support of life sciences. Narrowly targeted funding initiatives may well have value but are unlikely in isolation to generate the breadth of new knowledge required to lay the foundation for future cures.[18]

To say much the same by the analogy with which we began this book, we can see progress in science as a river, fed by multiple tributaries, whose eventual course cannot be predicted from study of the rivulets. It's a metaphor that Flexner also used: "Science, like the Mississippi, begins in a tiny rivulet in the distant forest. Gradually other streams swell its volume. And the roaring river that bursts the dikes is formed from countless sources."[19]

If we want the Loire to flow strong, to irrigate its vineyards in perpetuity and to be an artery that connects communities, if we want it to feed the imagination of artists and earn the love of the people, those tributaries must drain as extensive a landscape as possible, and be protected from industrial abstraction and from short-term overexploitation. It may be hard to see that any one rivulet matters much, but the consequences of damming their collective flow seems wholly predictable, even if it might take a generation for the consequences to be fully realized.

Curiosity-Driven Research

If you ask any academic scientist what it is that drives them, a common answer will be curiosity. There are exceptions, as in the single-mindedness of Schally and Guillemin in their pursuit of hypothalamic releasing factors or the desire of the pharmaceutical industry to develop specific therapies for specific conditions;

applied scientists in fields from computer science to clinical medicine have clear objectives. Nevertheless, it is worth pondering on the symbiotic relationship between applied and basic science to understand the place of curiosity-driven research.

The global pharmaceutical industry has vast resources at its disposal, including a veritable army of scientists. Their resources far exceed those available to academic biomedical science. Yet the pharmaceutical industry, in being monolithic and goal-oriented, depends on basic research conducted in academia. While there has been some outstanding fundamental research conducted by industry, industry has its limitations. Shareholders look askance at investment in basic science, when the returns from basic science cannot be corralled but may be realized, unpredictably, in any of many different sectors of the economy. The pharmaceutical industry has, therefore, strictly focused goals and works for returns realizable in as short a time as possible.

But in developing any drug, unanticipated effects arise. No pharmaceutical company however vast its resources can investigate all of these either for their potential utility or for potential adverse consequences. Instead, they call on the expertise that exists in the by-ways of academic science, where scientists, driven by curiosity, have developed expertise in answering questions of no visible utility. Collaborations between pharmaceutical industry and basic scientists arise generally when basic scientists have expertise that is absent from the industrial labs precisely because it has been developed in the absence of overt economic goals.

For academic scientists, such collaborations were not the purpose of their research, although they will be welcomed if the funding allows their curiosity fuller rein, and they will be welcomed as some visible justification of their efforts.

The shape of science as it grows is governed by diverse selection pressures—islands of discovery, seen as potentially important by any of many diverse understandings of importance, may flower by positive selection pressures. But negative selection is weaker—the cost of eliminating false trails is often greater than the worth of doing so. Progress in science is thus driven not by any self-correcting mechanism but simply arises due to its continuing expansion into new areas.

21 Lost in Citation

It is the unreasonable hope of every PhD student in biology that, however esoteric their studies, they will lead to something that transforms our understanding of a major disease. Such breakthroughs are not predictable: if they were, they wouldn't be breakthroughs. They typically come from curiosity-driven research, often in areas unrelated to their sphere of ultimate importance. This is well known by scientists, and it is not our point here. But unexpected findings, however potentially important, can go unrecognized unless some other things hold.

In 1974, Steve Hillier, a PhD student at the Tenovus Institute for Cancer Research in Cardiff, was developing a radioimmunoassay for testosterone, the male sex steroid hormone. The first step in developing a radioimmunoassay is to obtain an antibody that will recognize the hormone. In this case, it involved injecting rats with large amounts of testosterone conjugated to a large protein—bovine serum albumin (BSA)—to stimulate the production of antibodies to testosterone. Hillier began to wonder how the presence of large amounts of antibodies to testosterone would affect the rats—those antibodies should bind to any testosterone produced by the immunized rats and prevent it acting.

We might expect there to be many consequences, but a PhD student can't look at everything. There's a choice to be made, and Hillier's choice was prompted by a paper by Virendra Mahesh and Robert Goldblatt published in *Nature* in 1961. It reported that the ovaries of patients with Stein–Leventhal syndrome (now known as polycystic ovary syndrome) showed abnormal steroid synthesis.[1] Polycystic ovary syndrome is not rare: by some estimates, it affects between 10 percent and 20 percent of women of reproductive age; it is a leading cause of subfertility, and it has no cure. Mahesh and Goldblatt had proposed that the abnormalities of steroid synthesis might be the cause of the syndrome, but the paper had attracted little attention at the time—indeed, there was then comparatively little interest in the syndrome.

So, Hillier looked at the ovaries of female rats immunized against testosterone—and they all displayed polycystic ovaries. He published his findings in *Nature* in 1974.[2]

In May 2019, we searched the Web of Science for papers published on "polycystic ovaries and testosterone." The search returned a list of 3,653 articles that have been cited 115,937 times in 43,159 articles. The link between disrupted synthesis of steroid hormones in the ovary and polycystic ovary syndrome is now well accepted.

So, a young PhD student with a smart idea performs a clever experiment, publishes it in *Nature*, and the world beats a path to his door?

Not exactly. Hillier's paper has been cited just seventeen times. Just because you have a smart idea, do a clever experiment, and publish in *Nature*, don't assume that anyone will read it or care. Interest in testosterone and polycystic ovary syndrome did not take off until about 1995, twenty years after Hillier's *Nature* paper and more than thirty years after Mahesh and Goldblatt's paper.

Hillier can now recall his experience with a wry smile. Four things might have been important:

1. Hillier was unknown in the field—and so were his supervisors—he was working in a Cancer Research Institute, not with reproductive endocrinologists. You need to know and be known by the people you want to listen when you think you have something important to say.

2. The timing wasn't right. Louis Pasteur's aphorism *"Dans les champs de l'observation le hasard ne favorise que les esprits préparés."* (In the fields of observation, chance favors only the prepared mind) referred to the "prepared mind" of the researcher. But the minds of their readers also need to be prepared. If a tree falls in a forest where there is nobody to listen, does it make a sound?

3. Findings only have an impact when enough people have the interest and means to follow them up. Hillier wasn't able to follow up his own work, nor were his supervisors interested in doing so. Sometimes a paper is cited highly because of the twenty papers that followed that filled in the gaps, ironed out the flaws, and corrected the mistakes. If the authors don't follow up their own work, why should anyone else bother?

4. The title. Hillier's chosen title was: "Development of polycystic ovaries in rats actively immunized against T-3-BSA." Now Hillier winces. "T-3 BSA? That didn't mean anything to anybody. If only I'd just said *testosterone.*"

22 Conviction, Expectations, and Uncertainty in Science

> Every scientific statement must remain tentative forever. It may be corroborated, but every corroboration is relative to other statements which, again, are tentative. Only in our subjective experiences of conviction, in our subjective faith, can we be absolutely certain.
>
> —Karl Popper[1]

Scientific evidence is often inaccurate, always incomplete, and open to different interpretations. Many believe that the unique character, and ultimately the strength of Science, is not its certainty but its *uncertainty*, the incessant questioning of data, and the strife between conflicting interpretations that drives empirical attempts to resolve them.

Disputes and controversy are the stuff of scientific discourse, but resolving disagreements is seldom easy. Commonly, different scientists argue from different perspectives with no obvious way of resolving their arguments—they value some types of evidence or some sources of evidence more than others.

It would scarcely be surprising if scientists place most weight upon the evidence that they themselves have gathered and the methods that they are most familiar with. But social interests also affect how evidence is interpreted. Such influences include science funding priorities, journal publication priorities, and institutional policies; the pressures on individual scientists are diverse and pervasive.

Many sociologists of science have focused on the influence of ideologies, economics, and professional and personal interests. Basic scientists have often not taken seriously the notion that interests like these have much influence on their science. Such is the diversity of these factors, their variable and conflicting impacts and their overt conflict with the generally accepted norms of science, that basic scientists find it hard to see that, in the mixed and collaborative global community of scientists, social interests have much directional effect. In this, basic scientists have the particular advantage, if advantage it is, that they do not know where their findings or theories might lead to, beyond the immediate next step.

Science and Ideology

Some ideological influences are easy to recognize from historical studies of science. In Europe and America throughout the first half of the twentieth century, racist ideology fostered the rise of eugenics and underpinned the understanding of the nature of intelligence. This sad history has been thoroughly documented elsewhere, and we will not repeat it here.[2]

In Joseph Stalin's Russia, biology was distorted for a generation by the promotion of Ivan Michurin, who rejected the science of genetics in favor of a form of neo-Lamarckism. He believed that plants and animals could, by environmental pressures alone, adapt their forms and functions rapidly over just a few generations. This notion was taken up by Trofim Lysenko, Stalin's head of biology, who denied even the existence of genes.

Lysenko initiated a purge of biologists in the Soviet Union with a 1948 speech that portrayed those who favored a Mendelian understanding of genetics as "enemies of the people," and biological science in the Soviet Union, once strong, was crushed.

Lysenko claimed that plants could be heritably altered by graft hybridization and by controlling environmental conditions, and he was charged by Stalin with modernizing agriculture by implementing this.[3] From his "law of the life of species," he encouraged farmers to plant seeds close together, believing that saplings in thick clusters would sacrifice themselves, dying to ensure that future saplings would not be crowded, and he banned the use of fertilizers and pesticides. Catastrophically misguided application of his theories produced crop failures and famines that killed millions, leading eventually to his denunciation and removal in 1965.

Lysenko's theories had been supported by evidence that in subsequent years was suspected to be fraudulent, manipulated, or cherry-picked. He has been quoted as observing that "if you want a particular result you obtain it," and that "I need only people who will obtain what I require."[4]

It would seem scarcely worth stating that conclusions should be based on the objective merits of the evidence, rather than that evidence should be selected on the subjective merits of the conclusion, yet this is a principle widely neglected.

In retrospect it might seem that, while some of Lysenko's "evidence" was not credible, some was—and some biologists in the West took it seriously. These included J. B. S. Haldane, the leading evolutionary biologist of the time, and one who the Nobel laureate Peter Medawar is frequently cited as calling "the cleverest man I ever knew."[5] In 1940, Haldane pointed out that there were known nongenetic mechanisms by which certain acquired features of plants and animals could be inherited.[6] For example, since 1910, C. C. Little and his colleagues in Maine had bred different lines of mice, each with a characteristic liability to breast cancer.[7] In one line, 90 percent of females who survived for at least two years developed this disease, in another, only 5 percent did so. Members of the immune line did not become susceptible if they were caged with the susceptible line, so the liability seemed to be inherited. But if the young of the susceptible line were separated from their mothers at birth and fostered by immune females, they were much less likely to develop breast cancer. And this partial immunity was passed on to their children.

We now recognize that some agent is transmitted in the mice's milk that can cause breast cancer, and that this agent can be transmitted to future offspring either in utero or through their milk. In 1936, John Joseph Bittner identified such an "extrachromosomal" factor that could be removed from the milk by filtration, and which was later characterized by electron microscopy as a viral particle. The agent was shown to have reverse transcriptase activity like other retroviruses, such as that which causes AIDS, and the viral genome contains hormone-responsive elements that enhance viral replication during pregnancy.

Lysenko had vouched for four cases where tomatoes had been altered by grafting: evidence of transmissible agents that alter the shape of the fruit. Tomatoes are particularly susceptible to virus diseases and these diseases can be transmitted by grafting. But it is also possible to transmit viruses that immunize the plant against disease. To Haldane it seemed possible that the range of transmissible agents stretches from those that produce pathological effects, such as yellow patches on leaves, to others that affect growth.[8]

We now know that gene expression is governed by environmental factors in ways that can have effects throughout an individual's life—effects known as "developmental program-ming" and "early life programming." Some such effects can be transmitted through generations by partially understood mechanisms of "epigenetic inheritance" that arise not from modifications of the DNA sequences, but from biochemical modifications that subtly alter the level of expression of certain genes in particular tissues. These effects are still poorly under-stood; it is not known how epigenetic modifications are regu-lated or how any regulatory mechanisms might be encoded and inherited. These effects are likely to be quite unspecific—there is no sign of the kind of sophistication in the mechanisms that would be needed for fine adaptive regulation. These epigenetic changes can be passed (occasionally) from parents to offspring, but they are not permanent. Some might be "preadaptations"—where the phenotype of a newborn is "prepared" for features of the environment that it is born into as a consequence of effects of that environment on the parents. The scope of such inherited characteristics seems very limited, the mechanisms uncertain and unstable, and their adaptive value is more a matter of spec-ulation than something clearly established.

We cannot say that Lysenko was wrong in postulating nonge-netic modes of inheritance, or that he was wrong in asserting that some such mechanisms might in some circumstances benefit the progeny. We cannot say he was wrong in exploring whether nongenetic mechanisms of inheritance might be exploited for the benefits of society. We *can* say he was mistaken in denying the existence of genes, and therefore that he was wrong in believ-ing that the social and ideological interests of Western scientists had led, in this case, to false scientific conclusions. And we can say that he was fundamentally, deeply, and tragically wrong in favoring a conclusion because of its perceived merits rather than the objective merits of the totality of the evidence, and that he was utterly wrong in rejecting evidence simply because it did not conform to his preferred conclusions.[9]

Conflicts of Interest

Issues of conflict of interest arise in areas of applied science where financial self-interest is engaged. For example, many clinical trials are carried out to evaluate the effectiveness of drugs for different conditions. Many are funded by the manufacturers of the drugs—the companies cannot conduct such trials themselves; they need to persuade physicians to do so, physicians with established authority and who have access to patients. Some other trials are publicly funded—but generally for somewhat different reasons. Sometimes, they are needed to establish whether a drug treatment is cost-effective for example, or whether it is effective in particular patient groups. These may require designing a trial in a way that the manufacturers are reluctant to do.

Trials funded by industry are much more likely to produce results supportive of a drug than publicly funded trials, as has been catalogued by Ben Goldacre in *Bad Pharma*.[10] It would seem that the interests of commercial funders have thus often distorted the outcomes of clinical studies. That seems to be the view of Sir Iain Chalmers, who said of Goldacre's book, in a quote given on its cover: "This shocking book shows how people have suffered and died because the medical profession has allowed industry's interests to trump patient interests."

Chalmers was one of the founders in 1991 of the Cochrane Collaboration which was established to collate medical research findings and analyze them to inform evidence-based decisions about health interventions. Within twenty years, a network of 31,000 experts in 120 countries had collaborated on more than

5,000 reviews and meta-analyses published in the Cochrane Library. It is an organization globally respected for its high standards of rigor and objectivity, and it does not accept any financial support from profit-making sources.[11]

Goldacre opens *Bad Pharma* with an example of how industry funding compromises the evidence from drug trials:

> In 2010, three researchers from Harvard and Toronto found all the trials looking at five major classes of drug—[they] [. . .] then measured two key features: were they positive and were they funded by industry. They found over five hundred trials in total: 85 per cent of the industry-funded positive trials were positive but only 50 per cent of the government-funded trials were. That's a very significant difference.[12]

It's important to look carefully at evidence. The cited study looked at 546 trials and eventually included 220 industry-funded trials, of which 188 (85 percent) were positive.[13] They included thirty-six government-funded trials, of which eighteen (50 percent) were positive.

This is troublesome. There were only thirty-six government-funded trials in the study whose outcomes were known—and these were spread over five drug classes. Well, actually, no: none of the government-funded trials were on proton-pump inhibitors (drugs that reduce the production of acid by the stomach). Most were on just one of the five classes—antidepressants. Of the five classes, this had the *fewest* industry-funded trials. So, the study was not as big as it first seemed—220 industry-funded trials were ultimately compared with just thirty-six government-funded trials, and the industry-funded trials and government-funded trials were mainly concerned with different classes of drugs.

There were other differences too: the industry-funded trials were more likely to be multicenter trials—considered an important part of good trial design. Of the industry-funded trials, 97 percent were either on adults only or on children only, but more than 33 percent of the government-funded studies included both adults and children in their samples. Not only were the industry-funded trials and government-funded trials concerned with different classes of drugs, they seem to be quite different sorts of trials.

But there was another difference. In the industry-funded trials, the median sample size achieved in the final results was 191; in the government-funded trials, the median sample size was just twenty-three. Now that is what we must certainly call a very significant difference—in any of the different meanings of the word significant. Because we need to consider how a trial was classed as having a "positive" outcome:

> Results were considered favorable if they were statistically significant [. . .] and supported the efficacy or safety of the test drug or not favorable if they were not statistically significant for the efficacy or safety of the test drug.[14]

A trial was classed as "not favorable" not because the drug performed worse than its comparator or even because there was no difference in effect, but when there was no *significant* difference in effect. Statistical significance is very sensitive to sample size. The industry-funded studies were much better "powered"—because of their larger sample size they were better equipped to detect a true effect, if one existed.

So, far from providing evidence of something untoward about industry-funded research, this study doesn't show anything of the sort. Overall, the industry-funded trials were larger and of higher quality than the government-funded trials—something that is often found in comparisons of industry-funded trials and publicly funded trials—and not surprising given the disparity in resources available to industry and public funders.[15] But that's a conclusion of little relevance since the trials seem to be of different types and looking at different drugs.

Now, you, the reader might, at this point in our narrative, be asking in some consternation—what are we doing? Are we saying that Ben Goldacre's book is rubbish, that he got it all wrong?

We're not saying that, not at all. We have said nothing about the merits of Goldacre's general thesis nor of the particular case advanced in the first chapter, nor have we voiced any opinion agreeing or disagreeing with *any* of his conclusions. As it happens, when Gareth first read *Bad Pharma* he went out and bought ten copies to distribute to friends because the book had made an important case that he thought, and still thinks, deserves to be taken seriously. But it is not only homeopaths and the constellation of snake oil proponents that cite evidence selectively, that use straw man arguments to disparage their rivals, that use double standards in weighing evidence for and against their thesis.

What we have done here is stated that, at the opening of his book, Goldacre cited evidence as supporting his conclusion when a careful reading of that evidence shows that it is nothing of the sort. There is nothing surprising or unusual about this. People—and scientists, let us not forget, are still people—do it often. But taking a case seriously involves subjecting it to rigorous, forensic scrutiny in all its details.

It might be said that it is the duty of a scientist to have no friends to whom they have any loyalty; or, to put it a different way, it is their duty to friends and foes alike to find the flaws in their evidence and reasoning. This is not to disparage them, but to turn the cycle of science, the cycle of questions, evidence, and conclusions, around, to ask questions of the conclusions, to seek evidence to answer them, to draw better conclusions—better in the sense of informed by more and better evidence. Science does not end with conclusions that we call facts, the immutable stuff of dogma, but is an endless cycle of questioning things we might be tempted to treat as facts. It is not a cycle that spins on a point; it moves, albeit uncertainly, and its movement is what we call progress.

Open-Minded Thinking

Contemporary ideological influences are harder to recognize because they infiltrate not just scientific thinking but all thinking. For example, some see individuals as accountable for their own actions. They see the greatest good as best achieved through a combination of social authoritarianism and economic laissez-faire. Others see individual choices as so constrained by societal circumstances that there is little scope for free choice. They see no benefit in, as they see it, punishing the poor for being poor, and they see the greatest good as being achieved through a combination of social liberty and an equitable distribution of wealth. These mindsets are deeply embedded and affect the way that evidence is interpreted.

This is not the only problem. We commonly express complex and abstract ideas in ways that are suffused with metaphors and analogies; these make complex arguments easier to understand and give them an emotional salience that makes them memorable. In 2011, Paul Thibodeau and Lera Boroditsky looked at the influence of metaphors on how people drew conclusions from evidence. In one experiment, they looked at the effects of changing just one word. They asked subjects to read the following passage, which contained either the word "beast" or the word "virus":

Crime is a [beast/virus] ravaging the city of Addison. Five years ago, Addison was in good shape, with no obvious vulnerabilities. Unfortunately, in the past five years the city's defense systems have weakened, and the city has succumbed to crime. Today, there are more than 55,000 criminal incidents a year—up by more than 10,000 per year. There is a worry that if the city does not regain its strength soon, even more serious problems may start to develop.[16]

They then asked, "In your opinion, what does Addison need to do to reduce crime?" and "What is the role of a police officer in Addison?" The responses were analyzed to distinguish between whether they emphasized the police officer's role in (1) preventing crime or in (2) catching criminals.

The change of a single word made a difference: those who read the text framed by the word "virus" were much more likely to favor crime deterrence than those who read the text framed by the word "beast"—and this was a *bigger* difference than the predictable difference associated with their identified political leanings (Republican vs. Democrat).

In recent years, behavioral economists have identified a wide variety of *heuristics* (mental "shortcuts" that ease the cognitive load of making a decision) and *cognitive biases* (systematic errors in thinking) that influence our thinking and behavior in diverse contexts. The actively open-minded thinking scale was developed by psychologists to study how individuals differ in their styles of reasoning. It uses a questionnaire to gauge how willing subjects are to consider alternative opinions: their sensitivity to evidence that contradicts their current beliefs, their willingness to delay closure, and their disposition towards reflective thought (using items like: "If I think longer about a problem I will be more likely to solve it"). The scale is a good predictor of how individuals perform on tasks that engage heuristics and biases and of how often they fall into "reasoning traps" that identify superstitious thinking and belief in conspiracy theories.

Recently, the researchers who developed this scale noticed that several studies using a shortened form of the test had shown very high negative correlations with religiosity.[17] In itself, this did not seem surprising—a negative correlation had been consistently observed with the long version of the test—but the correlation was much stronger in studies that used the shortened test.

One particular type of item in the questionnaire accounted for these large correlations. These were statements such as "Beliefs should always be revised in response to new information or evidence." The researchers hadn't noticed that such items had a different salience for religious respondents than for secular respondents—and they hadn't noticed it because none of the research team were religious. For secular respondents, such items could be understood as being about a hypothetical belief, such as that one soft drink tasted better than another. But, for religious respondents, this was an item about their belief in God. Effectively, religious people and secular people were being evaluated using questionnaires that were worded identically but received differently: their responses could not be properly compared.

This had been overlooked because the researchers had *expected* that religiosity would be inversely associated with open-mindedness. They had not considered that, for religious people, spiritual belief was a special case of belief that was *not* incompatible with open-mindedness in other domains of belief, and in the short form of the test this was left unexplored.

Their paper is a brave acknowledgement of how their own confirmation biases had subverted an attempt to measure biases in reasoning and an account of how to adapt the scale to avoid this. They conclude, "Our own research on AOT becomes a case study of how ideological bias can infiltrate the work of even the most well-intentioned scientists."

There is an apparent belief amongst some sociologists that Science is best understood as guided by goals and interests. This sets the teeth of scientists on edge. It has been fostered by the current ethos of the sociology of science that it should aim to describe Science, not to judge it, and that it should seek to do so by passive observation, and not by understanding what scientists say or what they write but by identifying their goals and interests.

Yet, even in spheres of human activity where economic self-interest might be expected to be a dominant driver, such as in decisions about where and how to invest, decisions do not readily conform to any rationalist understanding. Decisions are subject to cognitive constraints—factors that affect what information we are able to use and how we are able to use it at any given time. Moreover, the interests of individuals are complex, multidimensional, malleable, and infinitely changeable. Put simply, the value that an individual places on achieving any given goal varies from individual to individual and changes with context and time.

23 Journals, Impact Factors, and Their Corrupting Influence on Science

Our easiest approach to a definition of any aspect of fiction is always by considering the sort of demand it makes on the reader. Curiosity for the story, human feelings and a sense of value for the characters, intelligence and memory for the plot.

—E. M. Forster[1]

The first scientific journals were mainly produced by scientific societies. They were conceived as the necessary means by which science was conducted, and often they struggled to be profitable. They were part of a common enterprise in which the interests of scientists and their institutions were bound by a common purpose. Scientists wrote papers, for which they were not paid; the papers were refereed by other scientists, who were not paid; and this process was managed by editors, who were not paid. The authors, referees, and editors then bought the published product at cost, either for themselves and their labs, or through the libraries of their institutions.

In the middle of the twentieth century, this curious economic model attracted the attention of commercial publishers who were not blind to the possibility of exploiting it for profit. In

1951, Robert Maxwell launched Pergamon Press and began to publish a series of new academic journals.[2]

Society journals were typically distributed at low cost to their members. They built a high reputation and had a high entry bar for publication: papers had to be complete and methodologically sound and they were expected to be authoritative. As such, authors seldom submitted papers that they did not believe had a reasonable expectation of being eventually accepted.

The new commercial journals took a different approach. The intent was to publish a journal that libraries would be obliged to subscribe to, even at eye-watering cost. One strategy was to expand a journal to the point where it published so much that it could not be overlooked, even if the quality of its content was suspect. In 1959, Pergamon owned forty journals; by 1965 it owned 150. Maxwell recognized the importance to his fledgling empire of earning the trust of scientists; he adopted the peer-review model, left the running of the journals to academics, and with that trust he was able to further expand his empire by striking deals with societies to publish journals on their behalf. He understood the commercial value of scientific reputation, and some journals in his portfolio like *Cell* and the *Lancet* became amongst the most prestigious of all.[3]

Maxwell's main concern was to expand: he saw that, with the discovery of the structure of DNA, there would be a boom in the life sciences that would lead to a proliferation of new fields, each of which could be served by its own journal. He branched out into psychology, social sciences, computer science—wherever he saw that science was expanding he saw the opportunity for a new journal, and science was expanding rapidly in all directions. There was increasing competition from other publishers; in particular, the Dutch publishing company Elsevier was expanding rapidly. But the competition didn't drive down prices: by 1988, a subscription to *Brain Research*—a Pergamon title that had become the largest journal in neuroscience—cost more than $5,000 (about $15,000 in 2019 cash terms). In 1991 Pergamon, with 400 journals in its portfolio, was sold to Elsevier for about $800 million (close to $1.5 billion today), making it the largest academic publisher in the world, and Elsevier raised the prices of the journals further. By 2017, Elsevier owned 2,500 journals and was publishing nearly half a million papers annually with an operating profit margin of 31 percent.[4]

The Importance of Prestige

Librarians became circumspect—increasingly their budgets were overspent, and choices had to be made. But in a predigital age, scientists relied on access to print journals through their institutional libraries. Nor could scientists readily find an alternative, cheaper way of publishing and disseminating their papers, because by now their own careers were intimately tied to the perceived prestige of the journals that they published in.

The new science of bibliometrics seemed to offer a solution to how rational choices about library subscriptions might be made. The influence of journals could apparently be measured by how often papers in them were cited. Thus, the *journal impact factor* was born, not as a measure of the quality of a journal, nor as a measure of the quality of the papers in it, but as an index of influence—a pragmatic metric for guiding the decisions of librarians.

However, the impact factor soon came to be understood to be a measure of quality. Several things followed: 1) there was pressure on journals to increase their impact factor; 2) individual scientists began to be judged by the impact factors of the journals that they published in; and 3) scientists began to focus their efforts on publishing in high-impact journals. The impact factor became a lever to oblige libraries to subscribe to particular journals.

Whereas Pergamon and Elsevier had focused on expansion first and developing prestige second, the Nature Publishing Group took the converse strategy—they concentrated on building prestige. *Nature*, first published in 1869, is one of the oldest scientific journals and built its reputation by wide recognition—it was a journal that had a very wide "popular" readership with a large base of individual subscribers. In the 1960s it could still be found on the magazine racks of high street bookshops in the United Kingdom. It brought science to the attention of a very broad academic and lay audience with short, clear, focused articles. One of the most famous is Crick and Watson's 1953 report of the structure of DNA—a model of clear, concise prose, in a single page, with its unforgettably understated conclusion: "It has not escaped our notice that the specific pairing we have postulated immediately suggests a possible copying mechanism for the genetic material."[5]

To publish an article in *Nature* was to announce a finding of potential importance. Scientists in the 1970s often dismissively spoke of *Nature* as a mere newspaper, but to publish there was a sought-after accolade. The "serious" publication of a finding, with full details, would follow in a "proper" journal.

But publishing a scientific magazine for a wide audience was not a safe or lucrative business. *Nature* soon turned to institutional subscriptions as its main source of revenue and to publishing longer and more densely technical papers. It gradually built its reputation as an elite journal, but still with its model of a professional editorial team. In 1970, *Nature* opened an office in Washington, DC, followed by offices in Paris, Tokyo, Munich, and Hong Kong.

In 2005 the *Nature* publishing group, now also publishing a series of *Nature Reviews* journals, merged with the German publishing group Springer Science + Business Media and with the Holtzbrinck Publishing Group. The new company, Springer Nature, used the *Nature* imprint to launch a proliferation of titles that by their name and from the moment of their launch were understood to be prestigious. They ignored the conventional wisdom that prestige was something that had to be earned by the patient building of a reputation. They also ignored the convention that journals should publish papers purely based on their scientific quality as judged in the process of expert peer review. Of papers submitted to *Nature*, more than 90 percent are never sent for peer review. These are rejected not by experts in the field, but by in-house editorial staff on their opinion as to whether the paper is likely to contribute to the journal's impact factor. Only the small percentage of submissions that pass this test are sent out for external review. It's a model now commonly operated by most journals that are "high-impact" or that aspire to such a status.

With this, the *Nature* group had begun appropriating, for its own commercial interests, the role of arbiter of importance and quality in science, a role that soon became prescriptive of how science should be done. This commercial takeover of the scientific enterprise occurred without the involvement of the established academic institutions of science and was scarcely even noticed by them until its implications became apparent. The takeover began by exploiting the impact factor.

The Impact Factor

The impact factor of a journal is a measure of how often, on average, papers published in that journal are cited, and it is calculated annually for every journal. For example, the 2018 impact factor of any particular journal is derived by dividing the number of citations made in 2018 to items that were published in that journal in 2017 and 2016, by the number of papers published in it in 2017 and 2016. So, if a journal published 100 papers in 2016 and 100 papers in 2017 and is cited 400 times in 2018, its 2018 impact factor will be 2.

Impact factors vary considerably between fields. In some fields, such as mathematics, papers typically have very few references, so impact factors will inevitably be low. Other fields are in the process of rapid growth: for these, a growing number of new papers will be citing a proportionately smaller source literature, and impact factors will be high.

Until recently, all journals were published in print and space was at a premium, so they restricted how many references could be included in a paper. More recent papers naturally have more citations for this reason alone. For some fields, there is a long-established body of older literature; by contrast, newly emerging fields have only recent papers to draw from. As citations to older papers have no effect on the impact factor, journals that covered long-established fields were disadvantaged.

To these disparities were added a wide variety of practices that distorted the measured impact factor.[6] Short items—letters, editorials, commentaries of papers, and news items—were sometimes cited but not counted as papers, so journals could inflate their impact factor by publishing more of them. Some editors wrote editorials that extensively cited papers published in their journal in the two preceding years, and in one extreme case, an annual editorial cited every paper published in the previous year. More commonly, journals began to apply subtle or overt pressure on authors to include references to papers published in that same journal. The most egregious of these manipulations were eliminated by the threat of sanctions on the journals concerned, but not all of these practices were indefensible. Editorials and commentaries provided a valuable way of contextualizing papers, and it seemed reasonable that where a published paper was apparently overlooked, an editor should draw an author's attention to it. It's hard not to have sympathy with editors who, perhaps aware of the idiocy of using impact factors as a measure of quality, used it cynically to their advantage.

However, journals had other, more pernicious ways of inflating their impact factors. Even within a field, some areas are cited more than others and journals began to prioritize these, lowering the bar on publication for papers that they thought might receive more citations. They began to prioritize citation rates above quality. For example, clinical guidelines were once published grudgingly as being of minimal perceived academic value but are highly cited for obvious reasons, and journals began to compete aggressively for the opportunity to publish them. Finally, editors noted that certain types of review article were often highly cited—unsurprisingly since it is the function of review articles to spread awareness of a field of science widely. This was especially true of review articles authored by "authorities" in a field. Journals that once accepted only primary research papers now began to actively solicit review articles, especially from the members of their editorial boards—members increasingly recruited not for any work that they were expected to do on behalf of the journal.

The Fallacies about Impact Factor

As journals promoted themselves by their impact factor, this added to the perception among authors that this was indeed a measure of quality. We should therefore reflect on three common fallacies:

1. The fallacy that "most papers published in high-impact journals are highly cited."

   Let us assume that a journal has the (very high) impact factor of 20 and that this is an "honest" impact factor, in the sense of being undistorted by publication of a proliferation of news items. Let us compare this with another journal with the more modest impact factor of 4—an impact factor common to many established journals of high reputation. We must remember the power law distribution of citations (chapter 16).

   In analyzing the oxytocin field (chapter 13), we noted that in each of five decades of research, about 20 percent of papers received about 60 percent of all citations, while 60 percent of the papers received just 20 percent of all citations. A similar relationship applies throughout science at every level—in every field, for every journal, and for the publications of every author.

   If a journal with an impact factor of 20 publishes 100 papers each year, then in the year in which it gains this impact factor it will have received 4,000 citations that are counted. By contrast, a journal of the same size with an impact factor of 4 will have received just 800 citations. The power-law distribution means that, for the high-impact journal, 60 percent of the papers (120 of them) will have received just 20 percent of those 4,000 citations; these will have been cited, on average, less than seven times each. Conversely, for the lower impact journal, 60 percent of the citations (480) will have been gathered by 20 percent of the papers—that is, by just forty of them. So, these 20 percent will have been cited, on average, about twelve times each—much more often than the average citation rate of 60 percent of papers in the high-impact journal.

The impact factor of *Nature* is inflated not only by the publication in *Nature* (and in their associated titles) of news items and commentaries but also by the same factors that cause the uneven distribution of citations. *Nature* papers are cited more often partly because they are more visible, partly because they are assumed to be authoritative, and partly by way of promoting the idea that a paper that cites them is addressing an important question.

Thus, the citations received by a paper are affected by where it is published—but perhaps this is simply because high-impact journals publish better papers. One way to test this is to see if the same paper is cited differently depending on where it is published. Almost all papers are published in only one journal—duplicate publication is regarded as serious misconduct when it is undeclared. However, there are exceptions: in medical journals, "white papers" are sometimes published simultaneously in several journals to maximize their reach. For example, in 2007, a white paper entitled "Clinical Trial Registration—Looking Back and Moving Ahead" was published in ten journals with impact factors ranging between less than 1 and 79. As first noted by Stuart Cantrill, the number of citations that each of these papers received (figure 23.1) correlates very strongly with the impact factor of the host journal ($r^2 = 0.9$).[7] Similar findings have been reported

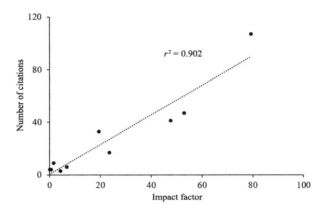

Figure 23.1

The relationship between the number of citations and journal impact factor for a single paper. "Clinical Trial Registration—Looking Back and Moving Ahead" by C. Laine, R. Horton, C. D. DeAngelis, J. M. Drazen, F. A. Frizelle, F. Godlee, C. Haug, et al., was published in 2007 in the *New England Journal of Medicine* 356: 2734–2736. It was also published in nine other journals. The straight line shows the linear regression. Citation data from the Web of Science cited reference search, May 2019. The impact factors are for 2018.

for other duplicated items.[8] Whatever the explanation, one thing is clear—the number of citations received by each paper can have said nothing at all about its quality, because these papers were all identical.

2. The fallacy that most highly cited papers are published in high-impact journals.

This follows from (1), given that far more papers are published in low-impact journals than in high-impact journals. In 2012, a study used the Web of Science to analyze citations to all papers published between 1902 and 2009, for all disciplines of the natural sciences and medical sciences together—about

30 million papers. The authors studied the relationship between how often papers were cited and the impact factor of the journal that they were published in.[9] Unsurprisingly these were significantly correlated—they could hardly not be. But the correlation was weak—the highest r^2 value was only about 0.3. In fact, the impact factor was a surprisingly poor predictor of how often any individual paper would be cited. The authors then identified the "top journals" as the top 10 percent by impact factor, and the "top papers" as the 10 percent with the most citations. In 1990, only 48 percent of the top articles had been published in the top journals. By 2009, this had fallen to 44 percent. Thus, papers published in high-impact journals are more likely to be highly cited, but most highly cited papers are not published in high-impact journals.

3. The fallacy that papers published in high-impact journals are more reliable than papers published in other, lower impact journals.

The evidence against this is that papers published in high-impact journals are more likely to be retracted than papers in lower impact journals and that most of these retractions are on the grounds of fraud.[10] This *might* be simply a consequence of the fact that these papers attract most scrutiny, but why it might be only to be expected is something we will come to later. However, several lines of evidence suggest that the methodological quality of studies is not any higher in journals of higher rank; on the contrary, it has been argued that "an accumulating body of evidence suggests the inverse: methodological quality and, consequently, reliability of published research works in several fields may be decreasing with increasing journal rank."[11]

It is sometimes assumed that high-impact journals have higher standards of peer review. This, according to chemists Antoinette Molinié and Geoffrey Bodenhausen, is a fallacy: "If asked to review a paper, we do not pay more attention when the request comes from *Science* than from the *Journal of Magnetic Resonance*. On the contrary! Since the likelihood that a paper actually will be accepted in *Science* appears slim, it is all too tempting for a referee to deliver a superficial review. Worse, the 'generalist' editors of nonspecialist journals do not know whom to ask. As a result, it is not rare to read papers in *Science* that are as muddled in their argumentation as spectacular in their claims. Such papers would never be accepted by the *Journal of Magnetic Resonance*! In fact, many articles that are published in *Science* are very ephemeral, while more fundamental long-lasting papers can only be found in the specialized literature."[12]

The Tyranny of Metrics

Despite its inadequacies, journal impact factor became a proxy for research quality, a metric commonly used by institutes, research groups, and grant and promotion committees to assess research and even individual researchers. As a consequence, researchers were incentivized to produce research that catered to the expectations of high-impact journals, producing a system that, in 2014, the Nobel laureate Sydney Brenner bluntly called "corrupt."[13]

In *The Tyranny of Metrics*, Jerry Muller listed some of the consequences of the inappropriate use of metrics.[14]

- **Misdescription of purpose:** in looking for quantifiable outcomes, the metric comes to be seen an as end in itself, and its intended purpose is neglected. For scientific journals as first conceived, their prime purpose was to publish original research of a high methodological quality. In seeking to publish only papers that will be highly cited, journals prioritize expected citation above quality, favoring areas of science and types of papers that have high expected citation rates. It might reasonably be thought that one purpose of a scientific journal is to disseminate the findings that it publishes and that a high impact factor is a measure of its success in this. But dissemination of findings is not itself a useful end, only the means to an end—an end that is well served only if what is disseminated is fit for the purpose of that end.

- **Goal displacement:** when a metric is used to judge performance, energy will be devoted to improving the scores, rather than to improving the outcomes for which the metric is presumed to be a proxy. Editorial boards have become preoccupied with improving their impact factor rather than trying to enhance the methodological and analytical rigor of the papers by improving the peer review process. When journals celebrate a high rejection rate as a supposed index of their selectivity and quality, they abandon the most important aspect of peer review—to provide constructive criticism and thereby to set goals for authors to improve their papers. *Nature* and *Science*, in rejecting most submissions without peer review, play little constructive part in the scientific enterprise.

In Muller's words, "Measurement is not an alternative to judgment; measurement *demands* judgment: judgment about whether to measure, what to measure, how to evaluate the significance of what's been measured, whether rewards and penalties will be attached to the results, and to whom to make the measurements available."[15]

The San Francisco Declaration on Research Assessment

The adverse consequences of the use of impact factors was unexpected and unintended, even by publishers, some of whom, including the editors of *Nature*, have scrambled to distance themselves from them.[16] In 2012, a group of editors and publishers met to develop the *San Francisco Declaration on Research Assessment* (DORA).[17] The declaration made a series of recommendations for funding agencies, institutions, publishers, organizations that produce research metrics, and individual researchers. By 2019 it had been signed by more than a thousand organizations including journals, societies, universities, and charities, and including all of the UK Research Councils, as well as more than 14,000 individuals.

The declaration promotes three themes:

1. "the need to eliminate the use of journal-based metrics, such as journal impact factors, in funding, appointment, and promotion considerations;

2. the need to assess research on its own merits rather than on the basis of the journal in which the research is published; and

3. the need to capitalize on the opportunities provided by online publication (such as relaxing unnecessary limits on the number of words, figures, and references in articles, and exploring new indicators of significance and impact)."

Plan S

It seems that to break the insidious presumptions that now permeate science, there might need to be a radical reconstruction of the system of scientific publishing. In 2018, a consortium of major research funders from ten European nations proposed a plan to do just that. The UK members of the consortium include the seven UK government Research Councils and the major UK research charity, the Wellcome Trust.

The plan, "Plan S," (as proposed in May 2019) will mandate that all of the scientists who they fund must publish all of their funded work in fully open-access journals, to make it freely available immediately on publication. The plan bars scientists from publishing in "hybrid journals"—journals that keep most content available only on subscription but allow authors to pay a fee that enables those paid-for articles to be freely available. The plan bars this on the (contested) grounds that such journals are in effect charging authors on the basis that their work will be openly available *and* charging libraries for making it available to their users. In its initial form, this rule in the plan would ban the funded scientists from publishing in most existing journals, including in *Nature* and *Science*.[18]

The logic of Plan S has two elements. First, it holds that governments and charities fund research as a public good and have an obligation to ensure that the funds provided in this way are not diverted for commercial interests to the detriment of that intended public good. Second, it holds that the ultimate payers—the citizens who pay taxes and the individuals who donate to charities—have a right to see what use their money has been put to.

In the United Kingdom, Plan S seems to allow no wiggle room for scientists to evade it. Universities are evaluated for their research excellence on a seven-year cycle. In the 2027 evaluation exercise, universities will only be able to submit papers for evaluation that comply with Plan S.

The Plan as proposed will entail a wholesale reconsideration of how research is evaluated, and its consequences are unforeseeable. There has been broad support across academia for open-access publishing, but there are some serious problems. The first is that the plan will undermine professional societies that depend on their own journals to fund their activities, despite society journals typically being rigorous and intimately attuned to the needs of the communities that they serve. Second, as most academic journals are published online, the operating costs are not necessarily very different between open-access journals and subscription journals—the difference is in how these costs are recovered. While subscription journals recover their costs (and their profits) efficiently from institutional libraries, open-access journals must recover theirs less efficiently from the authors (who must in turn recover them, if they can, from their institutions and funders).

Open-Access Journals

Subscription journals must, to gain their income, be valuable to their readers. Open-access journals, by contrast, must be valuable to the authors. This is an important difference. For a subscription journal, there is little direct cost in rejecting papers and some incentive to enhance the quality of the final published item. For open-access journals however, every paper rejected is a loss of income, and authors care more about acceptance and less about production quality. Accordingly, the danger is that the papers in open-access journals will be of lower quality through a lower bar on acceptance and also of lower production quality— and hence, paradoxically, *less* accessible through lower editorial standards. Bluntly, the danger is that open-access journals become vehicles for vanity publishing with minimal regard for readability.

Indeed, there is currently an epidemic of predatory open-access journals by new commercial publishers. Many of these have no competent peer review and have minimal production standards. They milk funds from naive authors, including many from developing countries who are trying to expand their research activity rapidly from a low funding base. Well-qualified experts are reluctant to spend time in reviewing papers for such journals.

In 2013, John Bohannon submitted fake articles to 304 open-access journals; the articles were concocted to include obvious, serious flaws.[19] By the time that Bohannon wrote his report, 157 of the journals had accepted the paper and only 98 had rejected it. Only 36 of the 304 submissions generated review comments that recognized any of the paper's scientific problems, and 16 of those papers were accepted by the editors despite the negative reviews.

Bohannon had submitted his fake articles mostly to online journals that any experienced scientist would probably have quickly recognized to be junk publications—but he did submit one to a notable exception, *PLoS ONE*, that responded with a comprehensive negative critique and rejected it. There are, now, many notable online journals such as those of the PLoS stable and *e-Life* (supported by the Wellcome Trust) that are seeking to deliver on the idealistic promise of open access while promoting scientific rigor and avoiding the distortions introduced by commercial journals. However, the number of junk journals is still growing rapidly, and they are a trap for the unwary.

Some open-access journals are run by prestigious scientific societies. For example, the Physiological Society and the American Physiological Society jointly operate *Physiological Reports*. This publishes papers that were initially submitted to their established subscription journals (such as the *Journal of Physiology*), but that were rejected not for any identified scientific weakness but on grounds of perceived impact—that is, on the perception that although sound, they will probably not be cited often. This practice panders to the distortions engendered by impact factor, but is perhaps a pragmatic way of moderating some of the worst consequences.

However, there are dangers. By rejecting the journal impact factor as a metric of any value at all in evaluating individual researchers or research groups, we are neglecting the wider issue of whether allocating resources for research on the basis of any bibliometrics is defensible.[20]

The Misuse of Bibliometrics

In 2010, Molinié and Bodenhausen, in an angry paper entitled "Bibliometrics as weapons of mass citation," noted that scientists are now often judged to be successful on the sole condition that their work has been abundantly cited.[21] This trend, by their analysis, appears to be supported not only by funding bodies, whose task is greatly simplified by these metrics, but also by scientists themselves whose "fascination for citation indexes often drives them beyond reason. Their obsession with their egos is much more perverse than the laziness of granting agencies. In some respects, scientists remind us of bodybuilders who, under the pretext of 'working out' in gyms and other *salons de musculation*, seem fascinated by their own bodies, reflected in floor-to-ceiling mirrors."

Richard Ernst, who was awarded the 1991 Nobel Prize in Chemistry for research that laid the foundations for nuclear magnetic resonance, proposed "remedies to save the dignity and creativity of scientists and researchers":

> Let us formulate a creed of scientists and researchers of all kind: Never ever use, quote, or even consult science citation indices! Let us appeal to the pride and honesty of researchers to derive their judgments exclusively by careful studies of the literature and other scientific evidence. It is better to refuse to comply with requests than to base your judgment on numeric bibliometric indicators! Let us incorporate this creed into our teaching, discrediting "number games" as incompatible with our goals of objectivity, credibility, fairness, and social responsibility, as researchers.[22]

In the next chapter, we consider how the current culture of journal publishing has had an effect on science.

24 The Narrative Fallacy: How a Good Story Beats Rigor and Balance

Once you have roughly defined your project, you have immediately to decide in which [high impact-factor] journal you would like to publish these future data. Once the choice is done, you look to the journal requirements, the way the published articles are constructed, what "stories" these papers generally tell, and what techniques are required to build a complete "story" if it has to be so.

Once you have made this complete analysis, because you don't master all the techniques to build your story, you negotiate collaborations, of course with the best scientists, if possible those who have already published in the journal chosen, to increase your chance of publishing!

Finally, you publish an article with 10–15 co-signatories, but for the [. . .] commissions that must judge the quality of your personal research, only the positions of first, second, or last matters! At the level of a research institute [. . .], only the number of such publications will reflect the excellence of research and, by a snowball effect, financial support and contracts will flow.

This well-intended advice was given by the head of one research institute to the head of a research group in another. It is quoted, anonymously but verbatim, from a letter to Gareth. The advice sits uneasily with the classic, apparently now obsolete idea that experiments need to be performed only when you don't know what their outcomes will be.

The importance of publishing in high-impact journals provides disincentives for scientists to engage in research that is unlikely to be published in them. This includes, as we have noted, studies that seek to test whether experiments with potentially important outcomes are repeatable. It also includes "descriptive" studies that are often a prerequisite for extending the scope of science into a new field.

For example, the brain is a complex structure with a huge number of distinct regions with unique populations of neurons; we know a lot about some of these—but almost nothing about most of them. To extend our knowledge into, say, the mammillary nucleus, about which little is known, would require a substantial amount of descriptive study. It would involve tracing the connections of the nerve cells, studying their intrinsic properties, identifying their biochemical features, before beginning to observe how these change in different physiological and experimental circumstances and before devising ways of manipulating their behavior or properties. This is novel science in the sense that we have no idea what we will find and so everything will be new. But it is hardly high-risk science: we can set out a feasible research program using established technology that will answer such questions. Perhaps the mammillary nucleus will turn out to be important; if so, the experiments that one day reveal that importance will depend on this kind of foundational detail. This detail will not be published in high-impact journals, because, at least at first, those papers will not be highly cited.

Nor will high-impact journals publish negative studies, those important studies that tell us what won't work. If a study asks a question to which we do not know the answer, why should we judge it to be good if it gives one answer but not worth publishing if it gives the other? High-impact journals will sigh in frustration—they are about impact, not quality, and negative studies and replication studies have little impact because they are seldom cited. But as we've observed, that they are seldom cited is a failing in the scientific literature, a failing that perpetuates misconceptions about the strength of evidence, and a failing that is exacerbated by the neglect of such studies by leading journals.

Nor will high-impact journals generally publish the anomalies that scientists encounter. Things that give rise to problems in interpretation, things that should give us pause for thought, are swept into a waste-paper basket or at best into a "pending file," of things that a scientist might come back to one day, if there is time or if something crops up to explain them.

However, the biggest problem with high-impact journals is not what they don't publish, but what they do publish, why they publish it, and the message that that sends to the scientific community.

Let's return to the quotation we began with to understand how well it captures what high-impact journals are about. It would be unsurprising to note that papers published in a high-impact journal tend to be written in a particular way, but this quotation expresses the sense that the studies that those papers report often conform to a formula. That should be a cause for some hesitation; we might expect that the nature of the studies should be determined by the nature of the question, not the nature of the journal.

Perhaps this is a misconception, so let us consider the rest of the quote. It advises forming a multidisciplinary collaboration with scientists who have published in that journal before. This may be good advice—and it is straight from the recipe book of how to spot a paper that will be highly cited. Multiauthored papers are more highly cited, including for one obvious reason—they have more authors to advertise them. Papers from well-known authors will also be more highly cited because of the assumed authority of the authors. Cumulative advantage in citation feeds on these things. A paper in a high-impact journal will be assumed to be good by those who have never read it. This assumption will be made on the basis that:

a) its authors must be good (because they have published in high-impact journals);

b) it was spoken about by invited speakers at important conferences (that is, by one or other of the authors, who had been invited to speak because they are known to be good); and

c) it was, after all, published in a high-impact journal.

We must consider what is implied by the suggestion that collaborations be negotiated. If such negotiations take the form "if you can show X" then "you can be an author of a paper that will be published in *Nature*" this would seem to be an intense incentive to show X. Intense incentives are dangerous, they incentivize the questionable practices that are hard to detect and which plague science.

We must also consider the emphasis on a story. As we have said, science is about communication; man is a story-telling animal. Simple stories that create a coherent explanatory chain and which are given salience by emotional cues in the telling of them are those most easily remembered and which will be most commonly repeated. But the problem with prioritizing the story in a scientific work is that it engages those very biases that we have recognized as subverting the "integrity" of science. Evidence that does not fit may be excluded from the account of the results or may be tortured by questionable analysis to make it appear to fit.[1] Experiments are designed to support the chosen narrative, not to challenge it rigorously. The discussion cites supporting evidence and ignores contradictory evidence. Bold claims are made of potential importance, and are allowed to go unchallenged because the emotional salience of a story is an essential element.

In papers that follow the advice with which we began this chapter, the elements of the paper's story will come from different methodologies and different laboratories. This has been defended on the grounds that these aspects combine to give the final product greater authority. Allegedly, there are at least two benefits here. First, the collaborative involvement between diverse authors, so it is thought, offers greater security against fraud, misrepresentation, and error. Second, by the principle of triangulation, when different independent sources of evidence converge on the same conclusion, our confidence in that conclusion should be greater.

Neither of these grounds is safe. Collaboration offers security in these terms when it is intimate, when different workers work closely and contiguously together. But when a collaboration is engaged to associate different expertise and methodologies, the scope for internal criticism is limited. When collaborations are not contiguous but remote, the scope for shared responsibility for integrity may be nonexistent.

The principle of triangulation has some value when a given problem is approached from independent directions. But this is not the case when different methodologies are used to construct links in a chain of evidence. A chain is only as strong as its weakest link; a link of straw is not stronger for being flanked by links of iron.

To assess the strength of a chain, every link must be tested. But reviewers of a paper that draws together multiple technologies will not have expertise in all of those elements; typically each reviewer will accept at face value the credibility of those with which he or she is not familiar, focusing only on those with which he or she *is* familiar. There will be several reviewers; where these have expertise in common, then the link that they examine may be robustly interrogated by their independent examinations. But where reviewers have similar expertise, they are also likely to have similar blind spots. Unsurprisingly, high-impact papers often contain elements that will be recognized by experts in that particular element as unsound when they see the paper in its published form.

Not everyone assumes that papers in high-impact journals are good. Every scientist can list a series of stinkers published in them, like the fraudulent vaccine study published in the *Lancet*, the fraudulent claims of cloning of human embryos in *Science*, and the seven *Nature* papers and eight *Science* papers on superconductivity that were fraudulently concocted.[2] High-impact journals have more cases of detected fraud; but whether this means that more of the work published in them is fraudulent is unclear. Perhaps there is a higher detection rate because of the greater attention paid to these papers.

Fraud is thought to be rare in science for a very human reason: scientists invest their sense of self-worth in the work they do, as any skilled artist does. If a scientist has, as they typically have, persevered through frustration to resolve an issue of some import to them. if they have incurred, as they typically have, a great expense of time, resources, and mental composure to bring their experiments to a final outcome—then, at the end, to just fiddle the numbers is not something done lightly. At least not for the sake of publishing in an "ordinary" journal—but for the chance to publish in *Nature* or *Science*?

In 2009, a meta-analysis of surveys that sought to estimate the prevalence of fraud reported that about 2 percent of scientists admitted having "fabricated, falsified or modified" data on at least one occasion, while about 14 percent reported having seen this in others at least once.[3] One driver of misconduct appears to be publication pressure—in 2019, one study of one particular area concluded: "publication pressure emerged as the strongest individual predictor of misconduct [. . .] the greater the publication pressure, the greater the reported misconduct."[4]

The main disincentive to fraud comes not from the risk of discovery, which may be low, but from the risk that the claim that was supported by the fraud will be proved false. Most claims made in papers never reach the kind of "fact-like" status from which they can be dislodged only with effort and difficulty. Some will just be ignored as unworthy of attention; some will be openly contradicted; many more will be quietly interrogated within a community, shown wanting, and forgotten. Within a small community, reputations are things built slowly, and ultimately center on "being right" or at least for being wrong only for very good reasons. For any individual, the best protection against the loss of their reputation is their integrity.

Much more common than fraud, and ultimately much more damaging, are the distortions of evidence and its interpretation that come under the broad heading of "questionable practices." Editors and reviewers seldom find reason to suspect fraud, but commonly detect misrepresentations of the evidence and gross statistical failings—but are likely to find these only when they arise in specific elements with which they are intimately familiar.

It is not therefore surprising that some papers published in high-impact journals are unreliable, or that this might be more often true of these than of papers published in other journals. Papers that follow a single methodology, published in specialist journals, are likely to be subject to close attention by referees expert in precisely the methodologies applied, and who know the field and its uncertainties in painful detail.

Postscript

The art of our necessities is strange
That can make vile things precious.
—William Shakespeare, *King Lear*

Journals joined the scientific body as symbiotes, evolved into parasites and corrupted it, and perhaps it is time to disinfect. If Plan S succeeds, it might bring down the commercial publishing empires of Elsevier and others. If it brings to earth the pretensions of *Nature* and *Science*, many scientists might be tempted to deny their staid reputations and dance in the streets. But while these journals thrive, the authority that they exert on the structure of science, an authority not granted to them voluntarily or on any rational basis, an authority assumed by them as an unintended consequence of commercial imperatives, holds the scientific enterprise in thrall.

Yet remedies sought in haste have unintended consequences, and if we bring down a corrupt edifice we must think about what will take its place. The strength of science is its diversity, a diversity that gives it the resources, in times of crisis, to engage in critical self-reflection and conceive of diverse alternative paths.

Well-intended remedies that seek to constrain that diversity need careful reflection.

Remedies that enhance diversity might be more effective, even if on a narrow view they might seem to be unfocused. Monopolies have potentially enormous power, but are relatively inflexible and prone to corruption, and it might not be a bad start to consider ways of dismantling the agents of monopolistic control of science. This might entail breaking the current model of scientific publishing and supporting a diversity of approaches to recognizing and enhancing quality in science.

Philosophers and sociologists of science have an important role in this debate. In recent times, they appear to have disengaged from the conduct of science; they speak about science, but seldom with scientists. Their insights, however acute, do not resonate amongst scientists because they do not concretely engage with scientists' own understanding of what they do and why they do it. Perhaps it is time for the flies to leave the walls of laboratories and break bread with the scientists that they observe.

25 Scholarship

Q209 Dr. Turner: When you were looking at the shape of the MRC scientific future did you do any sort of cost–benefit analysis—for want of a crude word—because clearly we are talking about investing large resources? Did you consider which kind of option would provide the best value for that investment in scientific terms for the future?

Professor Rothwell: The primary driver was the scientific value. Any discussions about costs were very much secondary and not in any detail at all. It was what would deliver the best science, which was first and foremost in all of the discussions.

Q210 Dr. Turner: Clearly one of your first considerations is in fact clinical links and the facilitation of translational research. What evidence do you have to show that actual co-location produces the best results?

Professor North: I think it would be quite difficult to move forward on the basis of evidence because if all science progressed on the basis of published evidence then progress in fact would be rather slow.[1]

The above exchange is from the Minutes of Evidence of the January 10, 2005, meeting of the UK Parliament's Select Committee on Science and Technology. Dr. Desmond Turner, for the Committee, was addressing Professors Nancy Rothwell and Alan North about the decision of the MRC to relocate the National Institute of Medical Research, a decision whose rationale was strongly contested by the institute staff.

The exchange captures a paradox. The MRC aspires to fund the best science, yet has no way of judging the quality of their science except by asking the scientists themselves—mainly, of course, those that they have funded. Yet here they were discounting the evidence of their own scientists at the National Institute who, it might be imagined, would best know where best to be located.

Such evidence as might have been used, bibliometric data for example, was apparently not used, perhaps because of its known failings. One clear message from scientometrics however is that the volume of science is directly related to the number of scientists, so any intervention that reduces the number of scientists that can be supported will inevitably affect scientific output. That seems not to have entered into any calculation by the MRC. This cost might be offset by gains in productivity, which can be measured, or quality, which might be measured by some criteria. But it seems that no attention was given to measuring productivity, or to defining criteria by which quality could be measured, or to how the evidence that already exists might inform their decision. No attention seems to have been given to planning how to subsequently assess the consequences of their decision.

Instead it appears, and this appearance might be completely mistaken, that the decision was made on the basis of the "gut instincts" of senior scientific advisors. The decision was understandable, if contestable, and perhaps there is no better way of coming to such a decision.

Yet the scientific attitude might demand that there should be some objective basis on which to take such decisions. One message from the study of science is that basic science has major economic benefits, but they come with a long lag and in unpredictable ways. Another is that the culture of science is important for its quality—the culture of skepticism, criticism, and integrity. A third is that the volume of highly influential papers is primarily related to the numbers of scientists who are active. Where that impact appears may be hard to predict, but it is easy to see that the more scientists who are supported, the greater the impact that will follow. Perhaps this can be finessed, that greater impact may be realized from fewer scientists, or that impact in particular areas might be made more likely. But we must start from the certainties, that the culture and volume of Science are key.

We have come to the end here of a journey in which we have documented some ways that Science and scientists are fallible. We observed that scientists are biased partisans in the promotion of their ideas and that this leads them to build arguments by cherry-picking evidence, flexibly interpreting evidence and data, and selectively citing sources. We observed how they are driven by nonsensical metrics and how commercial journals are subverting Science through practices that further undermine the rigor of Science. We observed how statistical tests have become rhetorical flourishes, disconnected from rational foundations. We observed that the scientific paper is, in some respects, a sham. Many research findings published in the recent literature may indeed be false.

Before we in indignation demand "What should we do about this?" and seek bold remedies, let us reflect that such thoughts often lead to cures far worse than the disease. For Science flourishes. Medicine has transformed our expectations for health and longevity. Science generally has delivered a world far, far better in most ways than any that people knew in any past times, and where those benefits aren't directly attributable to Science they are attributable in large part to the leisure, wealth, education, and health that Science has delivered. Our world has deep problems, some of which are attributable to the impact of Science, but it is to Science that we must look to solve them, for there is nowhere else to look.

In virtually every field of Science, progress is abundantly apparent. There are hiccups and reverses, nonsense happens in places for a while. It's not a universal monotonic progress—but the sophistication of our understanding deepens inexorably. So before contemplating the ills of Science and their possible cures, and knowing that any intervention always has unintended consequences, it's best that we think hard about what makes Science work.

The Passion of Science

Science is what scientists do and scientists are people, clever resourceful people, and what drives them is not the carrots of drug companies or the whips of funders, but their passion to understand. Of course, they are biased—but the biases come principally from that passion, the personal investment in their ideas. For every scientist, their research is deeply personal—they champion their favored positions, nurture them, and, sometimes, they fight for them.

It might be argued that science *ought* to be more dispassionate—that by emotional detachment we might prevent bias, avoid prolonged controversy, and make faster progress. This restates the old argument that divided reason from passion, that attributed order to the former and chaos to the latter, that set Apollo against Dionysus. This old argument is exactly the opposite of what the stories we have told suggest.

What if the success of science is deeply intertwined with our passion? Would Harris's ideas have been as rigorous without the challenge of Zuckerman? Would Trewavas have pursued his ideas with such rigor in the absence of an old guard resistant to change? Would a generation of researchers have been inspired without the impression that they were overcoming dogmatism?

Revolutionary science demands a faith, a faith that what we do is important, that it is interesting, and that it is doable. Most researchers will understand this—particularly at the start of their careers, when searching for the ideas worth dedicating their time to. In these early years, before they have accumulated the prestige that comes with publications and citations and before they are looked on kindly by funders, they will have to work long hours on speculative and vague ideas. Ideas that, often, only they see value in. They will need a powerful motivation to keep going when things go wrong, when they feel lost in the complexity of what they are doing, when they start to lose faith. They will have to learn to defend their ideas against the skepticism of elites. They will have to learn to convince others of the value that, for a time, only they saw. And they will succeed only by assembling their work into a persuasive new story about how things are and how they work.

Popper saw that science required imagination, bold ideas, and speculation, that it was only through these ideas that science could, in any sense, progress. Yet, Popper also saw that bold ideas required an *opposition*. The only way to have a strong opposition is to have people who are committed to their ideas, who are willing to defend them against attacks, and who are unwilling to abandon them at the first sign of trouble. Those who have seen fashions come and go, who have developed a sophisticated intuition that enables them to identify and dismiss the tides of nonsense that arise in their fields, and who defend their treasured ideas fiercely.

This can lead to blindness to alternative hypotheses, to skepticism of existing techniques, to denial of the worth of other research, and to the ignorance of findings on other sides of any debate. Our passion is a double-edged sword—but what is the alternative? Surely we want our researchers to care about what they do, to challenge established and new ideas, to defend themselves; above all, we want critical debate.

Passion is not enough. One cannot downplay the importance of scientists committing to intellectual integrity, however imperfectly this can play out in practice. As we have remarked, scientists likely deceive themselves rather than intentionally deceiving others. When evidence undermines their favored theories, when gaps in their understanding are pointed out, when the logical coherence of their positions falls apart, they must reflect on these and recognize when criticisms, painful as they may be, are valid.

Harris won the debate over Zuckerman because his arguments convinced others by demonstrating flaws in Zuckerman's arguments, but also because they pointed to new, unexplored areas of research. The same can be said for Trewavas. In both cases, these scientists constructed new stories, stories that were open-ended and expansive. They were constructed, championed, and defended by passionate individuals who inspired others to follow their lead.

Bold ideas often fail. The story of Theodosis and Catheline demonstrates the best of science—of determined and rigorous testing of ideas and of recognizing when we might be wrong. But recognizing when treasured stories have reached the end is always difficult. In the case of Steve Hillier, we saw that some bold ideas can fail to gain traction chiefly because they aren't championed, nurtured, or communicated effectively.

These qualities—reasonableness, passion, and honesty—are vital to our shared endeavor. We see these qualities in abundance, but we also see the many ways in which scientists diverge from ideal behavior. In constructing persuasive new stories, scientists can distort evidence and ignore previous work. The prevalence of citation bias, publication bias, and widespread use of inappropriate statistical tests in the biomedical literature points to a problem with how scientists are constructing evidence and utilizing previous research, potentially undermining the validity of knowledge claims, subverting reasoned debate and the soundness of advice.

Though these issues must be called out, we must remember that scientists are people—they are flawed, they are fallible, but, normally, they are trying their best. And, typically, these are not the failings of individuals but of social systems—of cascades of citations, of publication practices, of deficits in education, of flawed incentive structures, and of failures of collective responsibility.

Science Is Complex and in Perpetual Flux

Science is composed of a myriad of different, sometimes rival, methods, theories, and observations. This tripartite of the scientific enterprise—theory, data, method—exists in a complex symbiosis, one that evolves over time. As a result, how scientists describe and test their theories change over time, their methods of investigation evolve, even the meaning they ascribe to data changes. These are things that scientists themselves have a hand in deciding—nature does not decide for them. Popper alluded to this:

> We do not stumble upon our experiences, nor do we let them flow over us like a stream. Rather, we have to be active: we have to "make" our experiences. It is we who always formulate the questions to be put to nature [. . .] And in the end, it is again we who give the answer; it is we ourselves who, after severe scrutiny, decide upon the answer to the question which we put to nature.[2]

We cannot understand the differing meaning of terms like "oxytocin," "saturated fats," "intelligence," "function," or "stress" without recognizing that knowledge reflects what it is used for, as well as the history of questions, interests, and techniques that have built our current understanding.

Nor can we appreciate the unpredictability of the growth of our knowledge without this. Progress is very hard to predict in any specific domain of research. As we saw in the case of oxytocin research, a topic can move quickly into a domain of researchers with very different skills, background knowledge, and interests, and this can shape knowledge about that topic in unexpected ways.

Nor is the impact of a study predictable. As Greenberg showed, beliefs can emerge based on a biased selection of the available evidence and become reinforced through chains of citations. Citation bias is widespread—what is seen and known depends, in part, on what becomes popular, for whatever reason, in the scientific literature. Citation distortions tell us that, even when a study is well cited, what its audience believes it to have said can be very different from its intended message. The distortion of Jick and Porter's letter to support the use of opiates, the citations to Stang that converted the critical assessment of a methodology into somehow supporting its use, tell us that the meaning of studies can be corrupted in other papers. What is believed about a study, or group of studies, is not always established within them, but through chains of citations that reconstruct, selectively emphasize, or distort the original content.

Science is subject to group dynamics—to cascades of citations, exponential growth in publications, and collapses in output in particular topics—to fashions and fads. The replication crisis and dissemination bias pose problems for anyone who believes that the validity of Science is built on the robustness of its evidence base and the scrutiny of those who utilize it. But it is also a system that is fluid, that is self-governing and self-organizing in a way that makes it flexible to accommodate new interests, new techniques, new theories, and new observations. It enables ideas, data, and methods to spread rapidly across a community, but is also vulnerable to the amplification of error.

A Better-Directed Science?

The idea that science should be directed better is a natural one. In the United States, the allocation of biomedical resources is not closely aligned to the disease burden. This was described in a 2019 review as "a systemic misalignment between biomedical needs and resources," requiring "additional oversight, incentives, and feedback."[3] The authors argued that the choices of scientists about which direction to follow are shaped by a tension between tradition and innovation: scientists who adhere to a tradition may publish many papers that advance a focused research agenda, but this might limit their ability to "seize opportunities for staking out new ideas." High-risk innovation, they argued, is rare, because the reward does not compensate for the risk of failing to publish, and they proposed that, to alleviate this, funding agencies should proactively sponsor risky projects that test unexplored hypotheses.

We must consider what is meant by "innovation": innovation might be understood either as applying existing tools and understanding to a new area, or as developing new tools and a new understanding. To extend existing methodology into a new area might be worthwhile, but need not be risky. On the contrary, it is likely to lead to a ready flow of papers. Such research is risky only in the sense that nobody might take much notice; the new papers are likely to be poorly cited until that new area takes off. On the other hand, an important driver of science is the development of new methodologies. In neuroscience, these include ways of editing the genome of organisms to take experimental control of specific cells—the techniques of optogenetics and chemogenetics that bring cell activity under the control of light pulses or "designer drugs." Such innovations become necessary when the path to understanding appears blocked by the limitations of existing technology. For a researcher, the investment in innovation becomes feasible when enough funding is available to allow innovation to proceed in parallel with safer lines of research. New methodologies are often best developed with well-established "model systems" where questions are well defined and where there is a research community that will be quick to exploit them.

Risk Taking in Science

Research that is high-risk must, by definition, carry a high risk of failure. But scientists' careers depend on publication, as do their prospects of further grant funding—and so do the career prospects of the students and early researchers for whom they are responsible. Any strategy to fund high-risk research must protect these in some way. One way is to give young researchers who commit to high-risk research early tenure and guarantee continued funding of their research independent of their published output. It's a model once favored in leading universities but has now fallen fallow.

In 2014, Sydney Brenner reflected on the career of one of his former colleagues in Cambridge, Fred Sanger (1918–2013), who had won not one but two Nobel Prizes in Chemistry:

> A Fred Sanger would not survive today's world of science. With continuous reporting and appraisals, some committee would note that he published little of import between insulin in 1952 and his first paper on RNA sequencing in 1967 with another long gap until DNA sequencing in 1977. He would be labelled as unproductive, and his modest personal support would be denied. We no longer have a culture that allows individuals to embark on long-term—and what would be considered today extremely risky—projects.[4]

Problems That Scientists Have Made for Themselves

Many problems of Science are of scientists' own making; they have connived in the absurdity of evaluation by surrogate metrics, in the inflated respect for journal impact, and in the exaggeration of claims for impact. They have also connived in the centralization of research funding. This system gives most funding to those that have most, and by doing so makes scientists who have built their reputations at the bench become managers of scientists, something for which they are untrained and often unsuited. That process, when the careers of young scientists are fragile, makes some of the most innovative scientists strangely conservative. In protecting the careers of those they manage, they direct them to projects that, while superficially innovative, are yet disappointingly safe. It also makes them connive in the manufacture of high-impact papers that construct plausible narratives yet bypass skeptical review.

This is driven by the absurd insecurity of careers in Science; young scientists generally must go from one short-term position at one place to another somewhere else, and on and on. This trial by endurance leaves women behind more than men, gives scientists, randomly, experience of bad managers and good managers, of being fostered and of being exploited, of good practice and bad, of frustration and inspiration. The best thinkers often do their greatest work while young—Einstein was twenty-five years old in his *annus mirabilis* of 1905. Of the names mentioned here, Ayer was twenty-four when he wrote *Language Truth and Logic*, Popper thirty-two when he published *The Logic*, Latour thirty-two when he published *Laboratory Life*, Harris, Trewavas, Ioannidis, Price, and Kuhn all laid the foundations for their challenges to dogma in their thirties. To foster risk-taking and innovation we must accept that risks may succeed or fail for reasons unconnected with the qualities of the risk-taker, so risk taking should be backed by insurance for those who we ask to take those risks. And we would support the young who have least investment in the dogmas that demand fiercest scrutiny.

We don't support the young in this way for fear that science will become burdened with the consequences of poor choices, with ranks of the incompetent and unimaginative, the burned out and those whose technologies are obsolete. Science needs its innovators and iconoclasts, it skeptics and its dreamers, its technicians, its writers and poets, its advocates and administrators, even more, it needs its educators and students. And it needs managers with humility, who understand that it is a collective enterprise yet quintessentially individual, who understand that scientists are driven by passion more than by material gain, who seek material gain more as a surrogate of respect than for itself. Managers who understand that seeking all of these necessary attributes of Science in each scientist is foolish, and that their job is to sustain an ecology, not to grow orchids alone. In universities, where most Science is conducted, there are niches for many talents, and it is the role of managers to help individuals find them for the different stages of their careers.

Peer Review and Community

Science is not a monolith. Scientists communicate closely just with a few others—but could it be otherwise? They talk with those few whose work is most closely aligned to their own, those few who are most likely to see the real advances and the real problems. These are also those most likely to see past the rhetorical fluff that is understood to be part of the necessities of publishing in top journals and impressing those who won't read their papers in depth. These small communities may not always function effectively as scrutineers and gatekeepers of our understanding, but often they do, and in the long run we can hope that communities that have gone adrift will be embarrassed by exposure of their failings and fix themselves.

At the heart of a solution must be that feature which characterizes Science best—organized skepticism and "vigorous self-scrutiny." Many feel that the peer review process is broken. It has weaknesses, yet it is a cornerstone. Its manifest failings obscure its strength; the power of peer review to improve scientific papers is not mainly in the actual criticisms of referees but in the anticipated criticisms. Peer review has been described as organized humiliation.[5] Malicious, vituperative criticism is unacceptable, particularly when delivered anonymously. But this is rare—what is universally feared are the sharp and accurate criticisms that point to weaknesses of methodology or reasoning, that legitimately undermine a paper and thereby undermine the self-respect of the authors.

Scientists strive for the respect of their peers, and it is the anticipation of such potential humiliation that drives them to question themselves and their data and not present a manuscript to a journal unless and until they feel that it will withstand the fire of that criticism.

The criticisms of reviewers are seldom without foundation. When they are accurate, they stand on their own merits. If they are inaccurate, as they often are, they speak to the failure of the authors to make their case clearly. They often miss things; good reviewers strive hard to be thorough and constructive, but such is the volume of literature that reviewers who become known to be good can be overwhelmed with the numbers of papers and grants that they are asked to review and have to decline many.

Science could be better, it should be better, it was better perhaps when journals that were run by societies dominated, when their specialist journals held sway, with editors who knew the community, its flaws, and its potential intimately, and who understood the constructive importance of societies and their journals, and before impact factors poisoned the well. When communities were tight-knit and took collective responsibility for the quality of their output.

Could we, perhaps, just weed out poor science? How would we recognize it? Some might say we should shun low impact journals and sack poorly cited scientists. They should think before they speak and ask where the citations come from that give journals their impact factor and scientists their *h*-factor score. Overwhelmingly it is from low impact journals and poorly cited scientists. And why are these journals low impact and these scientists poorly cited? Not because they publish poor science and do poor experiments. They publish the papers that check, refine, extend, and test the findings in papers that receive high citations—these are the gatekeepers who sort the solid from the ephemeral, and their papers are often more trustworthy than papers in high-impact journals. In Science, it is not the first paper that really matters, but the hundreds that follow it. We celebrate that first paper; we grant it priority. But the first study is seldom the best, is often wrong, and often inflates the magnitude of the effect that it reports.

We earlier quoted Newton as saying, "If I have seen further it is by standing on the shoulders of Giants." Robert Merton, in a work of scholarship and wit, traced that quotation to the twelfth century and tracked its many variants.[6] He noted how it had begun as an aphorism of how a dwarf might see further by standing on the shoulders of a giant. He observed that some had asked how might a dwarf be raised there in the first place—and how others had noted that in science, a giant is only a giant because of the dwarves that have amassed on its shoulders, and how yet others had raised the prospect of the dwarves crushing the giant back into the ground. The citation giant, the "citation classic" paper, is a giant only because of the "dwarves" that have tested it, refined it, embellished it, corrected it, interpreted it and promoted it.

For all the misuse of bibliometrics we should find joy that the structure of science as revealed by citation networks tells stories not only of weakness but also of inspiration. Most papers that are highly cited do not come from the formula-factories of high-impact journals, and they rise because scientists themselves find them and propagate them. This democratization of influence gives science resilience—and it is the hubs and authorities in these networks that disseminate ideas beyond the small groups in which they incubate to wider communities. While the subsequent growth of any field is sporadic and unpredictable, these growth spurts, even when misguided, give science its vitality and invention. Big science may need large, directed teams of technician-scientists, but new ideas come mainly from small groups and small science.

The Importance of Scholarship

So what else is missing? Science also needs its philosophers and sociologists. Weighing and interpreting evidence requires *scholarship*, something we might define as a process of rigorous inquiry that advances the teaching, research, and practice of a given academic field. It requires some technical understanding of the methods used in a particular field and an appreciation of their strengths and limitations, as well as an appreciation of what is involved in a systematic search and meta-analysis. It also requires some philosophical understanding of the scientific methods of that field and an awareness of the social context of the science and how this impacts its evolution and its perceived importance. We have engaged with some features of the scientific literature that require particular awareness—including publication bias, selection bias, confirmation bias, and of some generic problems in science such as common issues in statistical analysis.

To combat these problems, we need scholars who spread good practice across their respective networks and who sound the alarm when things go wrong. John Ioannidis demonstrates that this can be done in respect of statistical weaknesses in Science. He has shaken the complacent, and his true impact will not be in his citations, massive though they are, but in the changes in education and scientific practice that will follow, driven mainly not by institutional directives but by the self-respect of scientists.

Scholarship was once considered to be a defining feature of universities, and education at a university was education in scholarship itself. Increasingly, scholarship has been neglected: as universities identify their primary roles as being in research and teaching; scholarship has come to be conceived as an activity in the "ivory towers" of academia, as though it were somehow disconnected from the real worlds of the research lab and the lecture theatre. However, scholarship is important both to the integrity of science, and to its effective application, as well as being still the essence of a good university education. If it is an activity conducted in ivory towers, protected thereby from the whims of politics and fashion, perhaps it is time to rebuild those towers.

Notes

Chapter 1: The Norms of Science, and Its Structure

1. Stephen Jay Gould, "Nurturing Nature," in *An Urchin in the Storm* (New York: W. W. Norton & Company, 1988), 150.

2. Gareth Leng, *The Heart of the Brain: The Hypothalamus and Its Hormones* (Cambridge, MA: MIT Press, 2018).

3. Thomas S. Kuhn, "Postscript," in *The Structure of Scientific Revolutions*, 2nd ed. (Chicago: University of Chicago Press, 1970 [1962]); Derek J. de Solla Price and Donald Beaver, "Collaboration in an Invisible College," *American Psychologist* 21 (1966): 1011–1018. See Kuhn's Postscript for extended discussion of the role of communities in scientific knowledge making.

4. Robert K. Merton, "The Normative Structure of Science," in *The Sociology of Science: Theoretical and Empirical Investigations* (Chicago: University of Chicago Press, 1973 [1942]).

5. Henry S. Pritchett, "Introduction," in *Medical Education in the United States and Canada*, ed. Abraham Flexner (New York: Carnegie Foundation for the Advancement of Teaching, 2002), xiii.

6. "Hansard (1976), Written Answers (Commons), Education and Science," HC Deb 29 March 1976 vol. 908 c371W, https://api.parliament.uk/historic-hansard/written-answers/1976/mar/29/school-leavers; P. Bolton,

"Education: Historical Statistics," *Commons Briefing Papers. Standard Note: SN/SG/4252* (2012).

7. Edward Sharpey-Schafer, "History of the Physiological Society During its First Fifty Years, 1876–1926: Part 1," *Journal of Physiology* 64, no. 3 (suppl.) (1927): 1–76.

8. Sharpey-Schafer, "History of the Physiological Society."

9. Sharpey-Schafer, "History of the Physiological Society."

10. Arthur Hugh Clough, "Say not the Struggle nought Availeth"

11. D. Fanelli, "How Many Scientists Fabricate and Falsify Research? A Systematic Review and Meta-analysis of Survey Data," *PLoS ONE* 4, no. 5 (2009): e5738.

Chapter 2: Popper and Kuhn, and Their Conceptions of What Science Is

1. Peter B. Medawar, *Advice to a Young Scientist*, Sloan Foundation Science Series (New York: Basic Books, 1981 [1979]), 39. Citations refer to the new edition.

2. D. L. Schacter, "The Seven Sins of Memory: Insights from Psychology and Cognitive Neuroscience," *American Psychologist* 54 (1999): 182–203.

3. Stephen Jay Gould, *An Urchin in the Storm: Essays about Books and Ideas* (New York: W. W. Norton, 1987), 150.

4. Karl Popper, *The Logic of Scientific Discovery* (London: Routledge, 2000 [1959]). An English translation of *Logik der Forschung. Zur Erkenntnistheorie der modernen Naturwissenschaft* (1934).

5. Karl Popper, *The Poverty of Historicism* (London: Routledge, 2002 [1957]), 124.

6. Popper, *The Logic of Scientific Discovery*, 4.

7. David Hume, *A Treatise of Human Nature*, Oxford Philosophical Texts, ed. D. F. Norton and M. J. Norton, (Oxford: Oxford University Press, 2002 [1738]).

8. Popper, *The Logic of Scientific Discovery*, 316.

9. Popper, *The Logic of Scientific Discovery*, 94.

10. An often-overlooked element of Popper's perspective is the focus he placed on scientific imagination: "Bold ideas, unjustified anticipations, and speculative thought, are our only means for interpreting nature: our only organon, our only instrument, for grasping her."—Popper, *The Logic of Scientific Discovery*, 280.

11. Kuhn, *The Structure of Scientific Revolutions*.

12. Armand M. Leroi, *The Lagoon: How Aristotle Invented Science* (London: Bloomsbury Publishing, 2014).

13. Thomas Aquinas, *Commentary on Aristotle's Physics* (New Haven, CT: Yale University Press, 1963), 136–137.

14. I. E. Drabkin, "Notes on the Laws of Motion in Aristotle," *American Journal of Philology* 59 (1938): 60–84.

15. C. Rovelli, "Aristotle's Physics: A Physicist's Look," *Journal of the American Philosophical Association* 1 (2015): 23–40.

16. Kuhn, *The Structure of Scientific Revolutions*, 16–17. "No natural history can be interpreted in the absence of at least some implicit body of intertwined theoretical and methodological belief that permits selection, evaluation, and criticism."

17. According to Kuhn (*The Structure of Scientific Revolutions*, 77–78), a scientific controversy arises when an existing paradigm encounters "anomalies," instances that contradict or cannot be explained in reference to the existing theoretical perspectives. However, Kuhn goes on to posit that the existence of anomalies by themselves is never enough to refute a scientific position, rather, advocates of a perspective will "devise numerous articulations and ad hoc modifications of their theory in order to eliminate any apparent conflict [. . .] though they may begin to lose faith and then to consider alternatives, they do not renounce the paradigm that has led them into crisis."

18. Kuhn (*The Structure of Scientific Revolutions*, 10, 77) argued "a scientific theory is declared invalid only if an alternate candidate is available

to take its place," and this alternative must be able to "attract an enduring group of adherents away from competing modes of scientific activity," while also being "sufficiently open-ended to leave all sorts of problems for the redefined group of practitioners to resolve," and capable of sufficiently explaining anomalies.

19. Kuhn, *The Structure of Scientific Revolutions*, 151, quoting Max Planck, *Scientific Autobiography and Other Papers*, trans. F. Gaynor (New York: Philosophical Library, 1949), 33–34.

Chapter 3: *Laboratory Life*

1. Paul Feyerabend, *Against Method: Outline of an Anarchistic Theory of Knowledge*, 4th ed. (New York: Verso Books, 2010), 43–44.

2. Feyerabend, *Against Method*.

3. Bruno Latour and Steve Woolgar, *Laboratory Life: The Construction of Scientific Facts* (Princeton, NJ: Princeton University Press, 1979).

4. Latour and Woolgar, *Laboratory Life*, 27–28.

5. Latour and Woolgar, *Laboratory Life*, 88.

Chapter 4: Is the Scientific Paper a Fraud?

1. T. E. Starzl "Peter Brian Medawar: Father of Transplantation," *Journal of the American College of Surgeons* 180: 332–336.

2. Peter B. Medawar, "Is the Scientific Paper a Fraud?" *The Listener* 70 (1963): 377–378.

3. S. M. Howitt and A. N. Wilson, "Revisiting 'Is the Scientific Paper a Fraud?'" *EMBO Reports* 15 (2014): 481–484.

4. D. T. Theodosis, D. A. Poulain, and J. D. Vincent, "Possible Morphological Bases for Synchronisation of Neuronal Firing in the Rat Supraoptic Nucleus During Lactation," *Neuroscience* 6 (1981): 919–929.

5. Literature search performed on Web of Science's Core Collection in 2018.

6. S. Monlezun, S. Ouali, D. A. Poulain, and D. T. Theodosis, "Polysialic Acid Is Required for Active Phases of Morphological Plasticity of Neurosecretory Axons and Their Glia," *Molecular and Cellular Neuroscience* 29 (2005): 516–524.

7. G. Catheline, B. Touquet, M. C. Lombard, D. A. Poulain, and D. T. Theodosis, "A Study of the Role of Neuro-glial Remodeling in the Oxytocin System at Lactation," *Neuroscience* 137 (2006): 309–316.

8. Catheline et al., "Neuro-glial Remodeling."

9. D. Fanelli, "'Positive' Results Increase down the Hierarchy of the Sciences," *PLoS ONE* 5, no.4 (2010): e10068, quotation on p. 5. See also D. Fanelli, "Negative Results Are Disappearing from Most Disciplines and Countries," *Scientometrics* 90 (2012): 891–904.

Chapter 5: The Birth of Neuroendocrinology and the "Stuff of Legend"

1. Bruno Latour and Steve Woolgar, *Laboratory Life: The Construction of Scientific Facts* (Princeton, NJ: Princeton University Press, 1979), 54

2. Geoffrey W. Harris, *Neural Control of the Pituitary Gland* (London: Edward Arnold, 1955).

3. Latour and Woolgar, *Laboratory Life*, 54.

4. F. L. Hisaw, "Development of the Graafian Follicle and Ovulation," *Physiological Reviews* 27 (1947): 95–119.

5. Harris, *Neural Control of the Pituitary Gland*.

6. Harris, *Neural Control of the Pituitary Gland*, 46.

7. Geoffrey W. Harris and D. Jacobsohn, "Functional grafts of the anterior pituitary gland," *Proceedings of the Royal Society of London B Biological Sciences* 139 (1952): 263–276.

8. Popper, *The Logic of Scientific Discovery*, 280.

9. Alun Chalfont, "Obituary: Lord Zuckerman," *Independent*, April 2, 1993, https://www.independent.co.uk/news/people/obituary-lord-zuckerman -1452840.html.

10. John Peyton, *Solly Zuckerman: A Scientist Out of the Ordinary*, (London: John Murray, 2001).

11. Frederick Dainton, "Lord Zuckerman (30 May 1904–1 April 1993)," *Proceedings of the American Philosophical Society* 139 (1995): 212–217.

12. S. Reichlin, "60 Years of Neuroendocrinology: Memoir: Working in the 'Huts' with the Professor: The First Maudsley Years," *Journal of Endocrinology* 226 (2015): E7–E11.

13. A. P. Thomson and S. Zuckerman, "The Effect of Pituitary-Stalk Section on Light-Induced Oestrus in Ferrets," *Proceedings of the Royal Society of London B* 142 (1954): 437–451.

14. B. T. Donovan and Geoffrey W. Harris, "Effect of Pituitary Stalk Section Light-Induced Oestrus in the Ferret," *Nature* 174 (1954): 503–504.

15. Reichlin, "60 Years of Neuroendocrinology."

16. A. P. Thomson and S. Zuckerman, "Anterior Pituitary Hypothalamic Relations in the Ferret," *Nature* 175 (1955): 74–76.

17. S. Zuckerman, "A Sceptical Neuroendocrinologist," in *Pioneers in Neuroendocrinology II*, ed. J. Meites, B. T. Donovan, and S. M. McCann (New York: Plenum Press, 1978), 403–411.

18. "To the community of neuroendocrinologists, Sir Solly's implacable pronouncements have been inexplicably equivalent to the rejection of Galileo's view of the relationship between the Sun and planet Earth." (Reichlin, "60 Years of Neuroendocrinology," E10.)

19. N. Wade, *The Nobel Duel: Two Scientists' 21-Year Race to Win the World's Most Coveted Prize* (Garden City, NY: Anchor Press/Doubleday, 1981).

20. B. A. Cross, "Chairman's introductory remarks," in *The Neurohypophysis: Structure, Function and Control*, ed. G. Leng and B. A. Cross, *Progress in Brain Research* 60 (1982): xi–xii.

21. Thomas S. Kuhn, "Postscript," in *The Structure of Scientific Revolutions*, 2nd ed. (Chicago: University of Chicago Press, 1970 [1962]), 10.

22. Kuhn, *The Structure of Scientific Revolutions*, 27.

23. Kuhn, 27.

24. Kuhn, 25.

25. National Center for Chronic Disease Prevention and Health Promotion, "Assisted Reproductive Technology 2015, National Summary Report," https://www.cdc.gov/art/pdf/2015-report/ART-2015-National-Summary-Report.pdf.

26. Kuhn, *The Structure of Scientific Revolutions*, 77.

27. Kuhn, 77.

28. N. Wade, "Guillemin and Schally: A Race Spurred by Rivalry," *Science* 200 (1978): 510–513.

29. R. Guillemin, "Peptides in the Brain: The New Endocrinology of the Neuron," *Science* 202 (1978): 390–402.

30. "Most intelligent people won't do isolation work . . ." Quoted in Wade, "Guillemin and Schally," 510–513.

31. Robert K. Merton, "Science and Democratic Social Structure," in *Social Theory and Social Structure*, enlarged revised ed. (New York: Free Press, 1968 [1957]), 605

32. Latour and Woolgar, *Laboratory Life*, 88

Chapter 6: The Language of Crisis and Controversy, and the Levers of Paradigm Change

1. S. Zuckerman, "Control of Pituitary Function," *Nature* 178 (1956): 442–443.

2. Zuckerman, "Control of Pituitary Function," 442–443.

3. A. J. Trewavas, "How Do Plant Growth Substances Work?" *Plant, Cell, and Environment* 4 (1981): 203–228.

4. Trewavas, "Plant Growth Substances," 203.

5. M. R. Knight, A. K. Campbell, S. M. Smith, and A. J. Trewavas, "Transgenic Plant Aequorin Reports the Effects of Touch and Cold-Shock and Elicitors on Cytoplasmic Calcium," *Nature* 352 (1991): 524–526.

6. Trewavas, "Plant Growth Substances."

7. A. J. Trewavas, "Profile of Anthony Trewavas," *Molecular Plant* 8 (2015): 345–351.

8. A. J. Trewavas and R. E. Cleland, "Is Plant Development Regulated by Changes in the Concentration of Growth Substances or by Changes in the Sensitivity to Growth Substances?" *Trends in Biochemical Sciences* 8 (1983): 354–357.

9. Knight et al., "Transgenic Plant Aequorin."

10. J. D. B. Weyers and N. W. Paterson, "Quantitative Assessment of Hormone Sensitivity Changes with Reference to Stomatal Responses to Abscisic Acid," in *Progress in Plant Growth Regulation. Current Plant Science and Biotechnology in Agriculture*, vol. 13, ed. C. M. Karssen, L. C. van Loon, and D. Vreugdenhil (Dordrecht, The Netherlands: Springer, 1992), 226–236.

Chapter 7: Logical Positivism

1. Voltaire, *Oeuvres complètes de Voltaire*, vol. 66, 418.

2. This was not a novel position: Hume (*A Treatise of Human Nature*, 35) argued that all of "the objects of human reason or enquiry may naturally be divided into two kinds: Relations of Ideas and Matters of Fact." The first kind included the mathematical sciences, whose propositions were "discoverable by the mere operation of thought." "Matters of fact" were those that could not be established by pure reason.

3. Alfred Jules Ayer, *Language, Truth, and Logic* (London: Penguin Books, 2001 [1936]), 18.

4. Quoted in Eric Pace, "A. J. Ayer Dead in Britain at 78; Philosopher of Logical Positivism," *New York Times*, June 29, 1989, https://nyti.ms/29yqeHX.

5. J. Passmore, "Logical Positivism," in *The Encyclopedia of Philosophy* 5, ed. P. Edwards (New York: Macmillan, 1967), 52–57.

6. Schlick, Moritz (1915), quoted in M. Friedman, "The Re-evaluation of Logical Positivism," *Journal of Philosophy* 88 (1991): 505–519.

7. H. Hans, Otto Neurath, and R. Carnap, "Wissenschaftliche Weltauffassung. Der Wiener Kreis" [The Scientific Conception of the World: The Vienna Circle], in *The Emergence of Logical Empiricism: from 1900 to the Vienna Circle*, ed. Sahotra Sarkar (New York: Garland Publishing, 1996 [1929]), 321–340.

8. Ayer, *Language, Truth, and Logic*, 18. Ayer, however, articulated this position to exemplify a view that he attributes to Moritz Schlick, but did not himself share this view.

9. Robert K. Merton, *The Sociology of Science: Theoretical and Empirical Investigations* (Chicago: University of Chicago Press, 1973), 270.

10. Otto Neurath, "Protocol Statements," in *Neurath Philosophical Papers 1913–1946: with a Bibliography of Neurath in English*, Vienna Circle Collection, ed. R. S. Cohen and M. Neurath (Dordrecht, The Netherlands: Springer, 1983 [1932]), 93. There were several different conceptualizations of the protocol sentence amongst the Vienna Circle's members. Neurath's own position was an attempt to resolve the problem he had identified with Schlick's approach (e.g. "here now red circle") and Carnap's approach (e.g. "I see a red circle here now"). To an extent, logical positivism began to fragment because Schlick, Carnap, and Neurath had conflicting perspectives on this issue. See Thomas E. Uebel, *Overcoming Logical Positivism from Within: The Emergence of Neurath's Naturalism in the Vienna Circle's Protocol Sentence Debate*, (Amsterdam: Rodopi, 1992).

11. Passmore, *Logical Positivism*.

12. This was the position of Wittgenstein in his later work on language-games: "For a large class of cases—though not for all—in which we employ the word meaning it can be explained thus: the meaning of a word is its use in the language." See Ludwig Wittgenstein, *Philosophical Investigations*, 4th ed., ed. P. M. S. Hacker and J. Schulte (Hoboken, NJ: Wiley-Blackwell, 2009 [1953]), 25. Note, however, this is *opposite* to that

which he had promoted in 1921 in his *Tractatus Logico-Philosophicus*, a work that heavily influenced the Vienna Circle and Ayer in particular.

13. C. G. Hempel, "Studies in the Logic of Confirmation (I.)," *Mind* 54 (1945): 1–26.

14. Ayer, *Language, Truth, and Logic*, 18.

15. Alfred Jules Ayer, interview by Bryan Magee, "Men of Ideas," BBC, 1978, https://youtu.be/4cnRJGs08hE.

Chapter 8: Ambiguity of Scientific Terms and the Construction of Meaning

1. E. M. Forster, *Aspects of the Novel* (London: Edward Arnold, 1927), 39.

2. A. J. Trewavas, "Green Plants as Intelligent Organisms," *Trends in Plant Sciences* 10 (2005): 413–419.

3. Trewavas, "Green Plants as Intelligent Organisms."

4. Gareth Leng, *The Heart of the Brain: the Hypothalamus and its Hormones* (Cambridge, MA: MIT Press, 2018).

5. Otto Neurath, "Protocol Statements," in *Neurath Philosophical Papers 1913–1946: with a Bibliography of Neurath in English*, Vienna Circle Collection, ed. R. S. Cohen and M. Neurath (Dordrecht, The Netherlands: Springer, 1983 [1932]), 92.

6. Richard P. Feynman, "New Textbooks for the 'New' Mathematics," *Engineering and Science* 28 (1965): 9–15.

7. The definition of myocardial infarction has changed over time. The current "Fourth Universal Definition of Myocardial Infarction (2018)" declares that "[Myocardial infarction] caused by atherothrombotic coronary artery disease (CAD) and usually precipitated by atherosclerotic plaque disruption (rupture or erosion) is designated as a type 1 [myocardial infarction]."

8. Medical Research Council, "Controlled Trial of Soya-bean Oil in Myocardial Infarction," *Lancet* 292 (1968): 693–700; P. Leren, "The Effect of Plasma-Cholesterol-Lowering Diet in Male Survivors of Myocardial

Infarction. A Controlled Clinical Trial," *Acta Medica Scandinavica Supplement* 466 (1966): 1–92.

9. K. Thygesen, J. S. Alpert, A. S. Jaffe, B. R. Chaitman, J. J. Bax, D. A. Morrow, et al. "Fourth Universal Definition of Myocardial Infarction (2018)," *Global Heart* 13 (2018): 305–338.

10. L. W. Kinsell, J. Partridge, L. Boling, S. Margen, and G. Michaels, "Dietary Modification of Serum Cholesterol and Phospholipid Levels," *Journal of Clinical Endocrinology* 12 (1952): 909–913.

11. A. Keys, J. T. Anderson, and F. Grande, "Prediction of Serum-Cholesterol Responses of Man to Changes in Fats in the Diet," *Lancet* 273 (1957): 959–966; E. H. Ahrens, W. Insull, R. Blomstrand, J. Hirsch, T. Tsaltas, and M. L. Peterson, "The Influence of Dietary Fats on Serum-Lipid Levels in Man," *Lancet* 272 (1957): 943–953.

12. H. L. Månsson, "Fatty Acids in Bovine Milk Fat," *Food & Nutrition Research* 52 (2008), https://doi.org/10.3402/fnr.v52i0.1821.

13. O. Paul, M. H. Lepper, W. H. Phelan, W. G. Dupertuis, A. Macmillan, H. Mckean, and H. Park, "A Longitudinal Study of Coronary Heart Disease," *Circulation* 28 (1963): 20–31; T. Gordon, A. Kagan, M. Garcia-Palmieri, W. B. Kannel, W. J. Zukel, J. Tillotson, et al. "Diet and Its Relation to Coronary Heart Disease and Death in Three Populations," *Circulation* 63 (1981): 500–515.

14. Select Committee on Nutrition and Human Needs, *Dietary Goals for the United States* (Washington, DC: US Government Printing Office, 1977).

15. R. Reiser, "Saturated Fat in the Diet and Serum Cholesterol Concentration: A Critical Examination of the Literature," *American Journal of Clinical Nutrition* 26 (1973): 524–555.

16. D. Schleifer, "The Perfect Solution: How Trans Fats Became the Healthy Replacement for Saturated Fats," *Technology and Culture* 53 (2012): 94–119.

17. Food and Nutrition Board, Institute of Medicine of the National Academies, *Dietary Reference Intakes for Energy, Carbohydrate, Fiber, Fat, Fatty Acids, Cholesterol, Protein, and Amino Acids (Macronutrients)* (Washington, DC: National Academies Press, 2005), 504.

18. A. D. Sokal, "Transgressing the Boundaries: Towards a Transformative Hermeneutics of Quantum Gravity," *Social Text* 46–47 (1995): 217–252. The hoax was revealed in A. D. Sokal, "A Physicist Experiments with Cultural Studies," *Lingua Franca* 6 (1996): 62–64. It was explained in A. D. Sokal, "Transgressing the Boundaries: An Afterword," *Dissent* 43 (1996): 93–99.

19. Harry M. Collins, "Stages in the Empirical Programme of Relativism," *Social Studies of Science* 11 (1981): 3–10.

20. Barry Barnes, "On Social Constructivist Accounts of the Natural Sciences," in *Knowledge and the World: Challenges Beyond the Science Wars,* The Frontiers Collection, ed. Martin Carrier, Johannes Roggenhofer, Günter Küppers, and Philippe Blanchard (Berlin: Springer, 2004).

21. For a reflection on the flaws of peer review see R. Smith, "Peer Review: A Flawed Process at the Heart of Science and Journals," *Journal of the Royal Society of Medicine* 99 (2006): 178–182. For a commentary on postpublication review, see G. I. Peterson, "Postpublication Peer Review: A Crucial Tool," *Science* 359 (2018): 1225–1226.

22. A. D. Sokal, "Commentary on Professor Barnes's Paper 'On Social Constructivist Accounts of the Natural Sciences,'" in *Knowledge and the World: Challenges Beyond the Science Wars*, ed. Martin Carrier, Johannes Roggenhofer, Günter Küppers, and Philippe Blanchard (Berlin: Springer-Verlag, 2004), 128–136.

23. Darien Graham-Smith, "The History of the Tube Map," *Londonist*, last modified April 6, 2018, https://londonist.com/2016/05/the-history -of-the-tube-map.

Chapter 9: The Totality of Evidence

1. H. L. Ho, "The Legal Concept of Evidence," *The Stanford Encyclopedia of Philosophy* (Winter 2015): https://plato.stanford.edu/archives/win2015/ entries/evidence-legal/.

2. A. E. Silverstone, P. Rosenbaum, F. Rosenbaum, R. S. Weinstock, S. M. Bartell, H. R. Foushee, C. Shelton, and M. Pavuk, "Polychlorinated Biphenyl (PCB) Exposure and Diabetes: Results from the Anniston

Community Health Survey," *Environmental Health Perspectives* 120 (2012): 727–732; S-L Wang, P-C Tsai, C-Y Yang, L. G. Yueliang, "Increased Risk of Diabetes and Polychlorinated Biphenyls and Dioxins; A 24-year Follow-up Study of the Yucheng Cohort," *Diabetes Care* 31 (2008): 1574–1579; M. Tang, K. Chen, F. Yang, W. Liu, "Exposure to Organochlorine Pollutants and Type 2 Diabetes: A Systematic Review and Meta-analysis," *PLoS ONE* 9, no. 10 (2014): e85556; P. Xun, K. He, "Fish Consumption and Incidence of Diabetes. Meta-analysis of Data from 438,000 Individuals in 12 Independent Prospective Cohorts with an Average 11-year Follow-up," *Diabetes Care* 35 (2012): 930–938; L. Marushka, X. Hu, M. Batal, T. Sadik, H. Schwartz, A. Ing, K. Fediuk, et al. "The Relationship between Persistent Organic Pollutants Exposure and Type 2 Diabetes among First Nations in Ontario and Manitoba, Canada: A Difference In Difference Analysis," *International Journal of Environmental Research and Public Health* 15, no. 3 (2018): E539.

3. Kyle Stanford, "Underdetermination of Scientific Theory," *The Stanford Encyclopedia of Philosophy* (Winter 2017): https://plato.stanford.edu/entries/scientific-underdetermination/.

4. Pierre Duhem, *The Aim and Structure of Physical Theory*, trans. P. W. Wiener (Princeton, NJ: Princeton University Press, 1954); Originally published as *La théorie physique: Son objet et sa structure* (Paris: Marcel Riviera & Cie, 1914). Quoted in Stanford, "Underdetermination of Scientific Theory."

5. Willard Van Orman Quine, "Two Dogmas of Empiricism," reprinted in *From a Logical Point of View*, 2nd ed. (Cambridge, MA: Harvard University Press, 1951), 20–46. Quoted in Marushka et al., "The Relationship between Persistent Organic Pollutants and Type 2 Diabetes."

6. N. D. Mermin, "The Science of Science: A Physicist Reads Barnes, Bloor, and Henry," *Social Studies of Science* 28 (1998): 603–662.

Chapter 10: Exaggerated Claims, Semantic Flexibility, and Nonsense

1. David Hume, *An Enquiry Concerning Human Understanding: And Other Writings*, Cambridge Texts in the History of Philosophy (Cambridge, England: Cambridge University Press, 2007), 101.

2. Ewan Birney, "Interviewee: Ewan Birney. How many genes are in the human genome?" *DNA Learning Centre*, https://www.dnalc.org/view/ 15295-How-many-genes-are-in-the-human-genome-Ewan-Birney.html.

3. M. Pertea and S. L. Salzberg, "Between a Chicken and a Grape: Estimating the Number of Human Genes," *Genome Biology* 11, no. 5 (2010): 206.

4. Pertea and Salzberg, "Between a Chicken and a Grape."

5. S. Ohno, "So Much 'Junk' DNA in Our Genome," in *Evolution of Genetic Systems,* Brookhaven Symposia in Biology, No. 23, ed. H. H. Smith (New York: Gordon and Breach, 1972), 366–370.

6. Ohno, "So Much 'Junk' DNA in Our Genome."

7. The Encyclopedia of DNA Elements (ENCODE) Consortium, "Project Overview," (2019), https://www.ensembl.org/info/website/tutorials/ encode.html.

8. A. G. Diehl and A. P. Boyle, "Deciphering ENCODE," *Trends in Genetics* 32 (2016): 238–249.

9. E. Pennisi, "ENCODE Project Writes Eulogy for Junk DNA," *Science* 337 (2012): 1159–1161.

10. D. Graur, Y. Zheng, N. Price, R. B. R. Azevedo, R. A. Zufall, and E. Elhaik, "On the Immortality of Television Sets: 'Function' in the Human Genome According to the Evolution-Free Gospel of ENCODE," *Genome Biology and Evolution* 5 (2013): 578–590.

11. S. R. Eddy, "The ENCODE Project: Missteps Overshadowing a Success," *Current Biology* 23 (2013): R259–261.

12. National Human Genome Research Institute, "The Cost of Sequencing a Human Genome," https://www.genome.gov/27565109/the-cost-of -sequencing-a-human-genome/. The cost cited here is less than the often quoted cost of $3 billion, which is the estimated cost of the entire project including many associated activities.

13. Genomics England, "The 100,000 Genomes Project" (2019), https:// www.genomicsengland.co.uk/about-genomics-england/the-100000 -genomes-project/.

14. "Pareidolia," Merriam-Webster's online dictionary, https://www.merriam-webster.com/dictionary/pareidolia.

15. Michel Gauquelin, *Written in The Stars* [*L'influence des astres*], trans. Hans Eysenck (New York: HarperCollins, 1988 [1955]).

16. P. Kurtz, M. Zelen, and G. Abell, "Results of the US Test of the 'Mars Effect' are Negative," *Skeptical Inquirer* 4, no. 2 (Winter 1979/80): 19–26; Claude Benski, *The "Mars Effect": A French Test of Over 1,000 Sports Champions* (Amherst, NY: Prometheus Books, 1996).

17. K. E. Peace, J. Yin, H. Rochani, S. Pandeya, and S. Young, "A Serious Flaw in Nutrition Epidemiology: A Meta-analysis Study," *International Journal of Biostatistics* 14, no. 2 (2018), 1.

18. Peace et al., "A Serious Flaw in Nutrition Epidemiology."

19. S. P. David, F. Naudet, J. Laude, J. Radua, P. Fusar-Poli, I. Chu, M. L. Stefanick, and J. P. A. Ioannidis, "Potential Reporting Bias in Neuroimaging Studies of Sex Differences," *Science Reports* 8, no. 1 (2018): 6082, https://doi.org/10.1038/s41598-018-23976-1.

20. C. Fine, "Is There Neurosexism in Functional Neuroimaging Investigations of Sex Differences?" *Neuroethics* 6 (2014): 369–409. See also C. Fine, "His Brain, Her Brain?" *Science* 346 (2014): 915–916.

21. David et al., "Potential Reporting Bias."

22. Barry Barnes, David Bloor, and John Henry, *Scientific Knowledge: A Sociological Analysis* (London: Athlone Press; Chicago: University of Chicago Press, 1996), 141.

23. J. Giles, "Statistical Flaw Trips Up Study of Bad Stats," *Nature* 443 (2006): 379.

24. Hans Eysenck and David Nias, *Astrology—Science or Superstition?* (CITY: Temple Smith, 1982); R. A. Crowe, "Astrology and the Scientific Method," *Psychological Reports* 67 (1990): 163–191; P. Hartmann, M. Reuter, and H. Nyborg, "The Relationship between Date of Birth and Individual Differences in Personality and General Intelligence: A Large-Scale Study," *Personality and Individual Differences* 40 (2006): 1349–1362.

25. Dorothy L. Sayers, *Gaudy Night: Lord Peter Wimsey, Book 12*, (London: Hodder Paperbacks, 1935), 391.

26. Leon J. Kamin, *The Science and Politics of IQ* (Potomac, MD: Lawrence Erlbaum Associates, 1974); W. H. Tucker, "Re-reconsidering Burt: Beyond a Reasonable Doubt," *Journal of the History of the Behavioral Sciences* 33 (1997): 145–162.

27. Nicholas Mackintosh, ed., *Cyril Burt: Fraud or Framed?* (New York: Oxford University Press, 1995)

28. A. J. Pelosi, "Personality and Fatal Diseases: Revisiting a Scientific Scandal," *Journal of Health Psychology* 24 (2019): 421–439.

29. D. F. Marks, "The Hans Eysenck Affair: Time to Correct the Scientific Record," *Journal of Health Psychology* 24 (2019): 409–420.

Chapter 11: Complexity and Its Problems for Causal Narratives

1. Edward N. Lorenz, "Deterministic Nonperiodic Flow," *Journal of Atmospheric Science* 20 (1963): 130–141.

2. Edward N. Lorenz, "Predictability: Does the Flap of a Butterfly's Wings in Brazil Set Off a Tornado in Texas?" (address given at the 139th Annual Meeting of the American Association for the Advancement of Science, Boston, MA, December 29, 1972).

3. Gareth Leng, "Butterflies, Grasshoppers, and Editors," *Journal of Neuroendocrinology* 16 (2004): 1–2.

4. W. S. Franklin, "Review of P. Duhem, *Traité Elementaire de Méchanique Chimique fondée sur la Thermodynamique*, Two Volumes. Paris, 1897," *Physical Review* 6 (1898): 170–175.

5. Thomas Hobbes, *De Cive [On the Citizen]* (London: Printed by J.C. for R. Royston, at the Angel in Ivie-Lane, 1651), 4. A version can be found online here: http://www.public-library.uk/ebooks/27/57.pdf.

6. Quoted in Lee Dye, "Nobel Physicist R. P. Feynman of Caltech Dies," *Los Angeles Times*, February 16, 1988, https://www.latimes.com/archives/la-xpm-1988-02-16-mn-42968-story.html.

Chapter 12: Publication and Citation

1. Michel Foucault, *Archaeology of Knowledge,* 2nd ed. (London: Routledge, 1969), 26–27.

2. A. F. J. Van Raan, "Sleeping Beauties in Science," *Scientometrics* 59 (2004): 467–472; Derek J. de Solla Price, "Networks of Scientific Papers," *Science* 149 (1965): 510–515.

3. M. V. Simkin and V. P. Roychowdhury, "Read Before You Cite!" *Complex Systems* 14 (2003): 269–274.

4. Henry H. Bauer, *Scientific Literacy and the Myth of the Scientific Method* (Chicago: University of Illinois Press, 1992).

5. Logan Wilson, *The Academic Man: A Study in the Sociology of a Profession* (New Brunswick, NJ: Transaction Publishers, 1995 [1942]), 195.

6. Karl Popper, *The Open Society and Its Enemies* (Princeton, NJ: Princeton University Press, 2013 [1945]), 426.

7. E. M. Forster, *Aspects of the Novel,* first electronic edition (New York: Rosetta Books, 2002), 72.

8. J. R. Cole and S. Cole, *Social Stratification in Science* (Chicago: University of Chicago Press, 1973).

9. L. Bornmann and H. D. Daniel, "What Do Citation Counts Measure? A Review of Studies on Citing Behavior," *Journal of Documentation* 61 (2006): 45–80.

10. G. N. Gilbert, "Referencing as Persuasion," *Social Studies of Science* 7 (1977): 113–122; Bruno Latour, *Science in Action: How to Follow Scientists and Engineers through Society* (Milton Keynes, UK: Open University Press, 1987); D. Edge, "Quantitative Measures of Communication in Science: A Critical Overview," *History of Science* 17 (1979): 102–134.

11. G. Wolfgang, T. Bart and S. Balázs, "A Bibliometric Approach to the Role of Author Self-citations in Scientific Communication," *Scientometrics* 59 (2004), 63–77.

12. M. W. King, C. T. Bergstrom, S. J. Correll, J. Jacquet, and J. D. West, "Men Set Their Own Cites High: Gender and Self-citation across Fields and over Time," *Socius* 3 (2017): 1–22.

13. B. Johnson and C. Oppenheim, "How Socially Connected Are Citers to Those That They Cite?" *Journal of Documentation* 63 (2007): 609–637; M. Wallace, V. Larivière, and Y. Gingras, "A Small World of Citations? The Influence of Collaboration Networks on Citation Practices," *PLoS ONE* 7 (2012): e33339, https://doi.org/10.1371/journal.pone.0033339.

14. Robert K. Merton, "Priorities in Scientific Discovery," *American Sociological Review* 22 (1957): 635–659.

15. Barry Barnes, David Bloor, and John Henry, "Words and the World," in *Scientific Knowledge: A Sociological Analysis* (London: Athlone Press; Chicago: University of Chicago Press, 1996), 46–80.

16. Gareth Leng, *The Heart of the Brain: the Hypothalamus and Its Hormones* (Cambridge, MA: MIT Press, 2018).

17. However, this is field dependent and appears to be a trend of the average age of references becoming somewhat older in recent times. See V. Larivière, É Archambault, and Y Gingras, "Long-Term Variations in the Aging of Scientific Literature: From Exponential Growth to Steady-State Science (1900–2004)," *Journal of the American Society of Information Science* 59 (2008): 288–296; P. D. B. Parolo, R. K. Pan, R. Ghosh, B. A. Huberman, K. Kaski, and S. Fortunato, "Attention Decay in Science," *Journal of Infometrics* 9 (2015): 734–745.

18. C. Chorus and L. Waltman, "A Large-Scale Analysis of Impact Factor Biased Journal Self-citations," *PLoS ONE* 11, no. 8 (2016): 0161021, https://doi.org/10.1371/journal.pone.0161021.

19. Robert K. Merton, "The Matthew Effect in science," *Science* 159 (1968): 56–63; Derek J. de Solla Price, "A general theory of bibliometric and other cumulative advantage processes," *Journal of the American Society of Information Science* 27 (1976): 292–306.

20. M. Callaham, R. L. Wears, E. Weber, "Journal Prestige, Publication Bias, and Other Characteristics Associated with Citation of Pub-

lished Studies in Peer-Reviewed Journals," *JAMA* 287, no. 21 (2002): 2847–2850.

21. Popper, *The Open Society and Its Enemies*, 426.

22. I. Newton, "Letter from Sir Isaac Newton to Robert Hooke," Historical Society of Pennsylvania, *1675*, https://discover.hsp.org/Record/dc-9792/Description.

23. Eugene Garfield, *Citation Indexing: Its Theory and Application in Science, Technology, and Humanities* (New York: Wiley, 1979), 23–24.

24. Derek J. de Solla Price, *Little Science, Big Science . . . and Beyond* (New York: Columbia University Press, 1986 [1963]); E. Garfield, I. H. Sher, R. J. Torpie, *The Use of Citation Data in Writing the History of Science* (Philadelphia: Institute for Scientific Information, 1964); E. Garfield, "Citation Indexes for Science," *Science* 122 (1955): 108–111; M. M. Kessler, "Comparison of the Results of Bibliographic Coupling and Analytic Subject Indexing," *American Documentation* 16 (1965): 223–233.

25. B. Cronin, "A Hundred Million Acts of Whimsy?" *Current Science* 89 (2005): 1505–1509.

26. Bruno Latour, *Science in Action: How to Follow Scientists and Engineers through Society* (Milton Keynes, UK: Open University Press, 1987), 37–38.

27. Latour, *Science in Action*, 40.

28. Price, *Little Science, Big Science*, xv.

29. Price, "Quantitative Measures of the Development of Science," *Archives Internationales d'Histoire des Sciences* 4 (1951): 86–93.

30. L. Bornmann and R. Mutz, "Growth Rates of Modern Science: A Bibliometric Analysis Based On the Number of Publications and Cited References," *Journal of the Association of Information Science and Technology* 66 (2015): 2215–2222.

31. D. Fanelli and V. Larivière, "Researchers' Individual Publication Rate Has Not Increased in a Century," *PLoS ONE* 11 (2016): e0149504.

32. Price, "Networks of Scientific Papers."

33. P. Albarrán, J. A. Crespo, I. Ortuño, and J. Ruiz-Castillo, "The Skewness Of Science in 219 Sub-fields and a Number of Aggregates," *Scientometrics* 88 (2011): 385–397; Michal Brzezinski, "Power Laws in Citation Distributions: Evidence from Scopus," *Scientometrics* 103 (2015): 213–228; M. Golosovsky, "Power-Law Citation Distributions Are Not Scale-Free," *Physical Review E* 96 (2017): 032306.

34. Price, "Bibliometric and Other Cumulative Advantage Processes."

35. Merton, "The Matthew Effect in Science," 58.

36. Merton, 59.

37. Price, "Bibliometric and Other Cumulative Advantage Processes," 293.

38. Price, 293.

Chapter 13: A Case Study of a Field in Evolution

1. Ludwik Fleck, *Genesis and Development of a Scientific Fact* (Chicago: University of Chicago Press, 1979). Originally published as *Enstehung und Entwicklung einer wissenschaftlichen Tatsache: Einführung in die Lehre vom Denkstil und Denkkollektiv* (Basel: Benno Schwabe & Co., 1935).

2. J. S. Liu, L. Y. Y. Lu, "An Integrated Approach for Main Path Analysis: Development of the Hirsch Index as an Example," *Journal of the American Society for Information Science* 63 (2012): 528–542.

3. Web of Science is a collection of databases. We used the Core Collection to locate articles that had "oxytocin" in the title, and from these selected primary research papers published in English in international, peer-reviewed academic journals. Results were filtered to exclude any paper not identified as such by Web of Science. The first search was conducted in May 2018 and analyzed by Gareth manually. A second search was conducted in May 2019 to collect additional distribution data.

4. H. H. Dale, "On Some Physiological Actions of Ergot," *Journal of Physiology* 34 (1906): 163–206. Ergot is the product of a small fungus, *Claviceps purpurea*, which grows on rye. Like many herbal remedies, ergot is

not benign: see J. W. Bennett and R. Bentley, "Pride and Prejudice: The Story of Ergot," *Perspectives in Biology and Medicine* 42 (1999): 333–355.

5. W. B. Bell, "The Pituitary Body and the Therapeutic Value of the Infundibular Extract in Shock, Uterine Atony, and Intestinal Paresis," *British Medical Journal* 2 (1909): 1609–1613.

6. B. P. Watson, "Pituitary Extract in Obstetrical Practice," *Canadian Medical Association Journal* 3 (1913): 739–758.

7. O. Kamm, "The Dialysis of Pituitary Extracts," *Science* 67 (1928): 199–200. Pitocin is a trademark, now used by Parke-Davis for synthetic oxytocin.

8. D. Llewellyn-Jones, "The Place of Oxytocin in Labour," *British Medical Journal* 2 (1955): 1364–1366.

9. M. Bodanszky and V. Du Vigneaud, "A Method of Synthesis of Long Peptide Chains Using a Synthesis of Oxytocin as an Example," *Journal of the American Chemical Society* 81 (1959): 5688–5691.

10. G. Leng, R. Pineda, N. Sabatier, and M. Ludwig, "60 Years of Neuroendocrinology: The Posterior Pituitary, from Geoffrey Harris to Our Present Understanding," *Journal of Endocrinology* 226 (2015): T173–185.

11. D. B. Hope, V. V. Murti, and V. Du Vigneaud, "A Highly Potent Analogue of Oxytocin, Desamino-oxytocin," *Journal of Biological Chemistry* 237 (1962): 1563–1566.

12. R. M. Buijs, "Intra- and Extrahypothalamic Vasopressin and Oxytocin Pathways in the Rat. Pathways to the Limbic System, Medulla Oblongata, and Spinal Cord," *Cell and Tissue Research* 192 (1978): 423–435.

13. Buijs, "Vasopressin and Oxytocin Pathways in the Rat."

14. J. B. Wakerley and D. W. Lincoln, "The Milk-Ejection Reflex of the Rat: A 20-to 40-fold Acceleration in the Firing of Paraventricular Neurones During Oxytocin Release," *Journal of Endocrinology* 57 (1973): 477–493.

15. C. A. Pedersen, J. A. Ascher, Y. L. Monroe, and A. J. Prange Jr., "Oxytocin Induces Maternal Behavior in Virgin Female Rats," *Science* 216 (1982): 648–650.

16. K. M. Kendrick, E. B. Keverne, and B. A. Baldwin, "Intracerebroventricular Oxytocin Stimulates Maternal Behaviour in the Sheep," *Neuroendocrinology* 46 (1987): 56–61.

17. T. R. Insel, "Oxytocin and the Neurobiology of Attachment," *Behavioral Brain Sciences* 15 (1992): 515–516.

18. M. Kosfeld, M. Heinrichs, P. J. Zak, U. Fischbacher, and E. Fehr, "Oxytocin Increases Trust in Humans," *Nature* 435 (2005): 673–676.

19. R. T. Hoover, "Intranasal Oxytocin in Eighteen Hundred Patients. A Study on Its Safety as Used in a Community Hospital," *American Journal of Obstetrics and Gynecology* 110 (1971): 788–794.

20. For figure 5 we looked only at citations to primary research papers—we excluded reviews, which generally are more highly cited and which have become increasingly common.

21. C. F. Blanford, "Impact Factors, Citation Distributions, and Journal Stratification," *Journal of Materials Science* 51 (2016): 10319–10322.

Chapter 14: Where Are the Facts?

1. Ludwik Fleck, *Genesis and Development of a Scientific Fact,* (Chicago: University of Chicago Press, 1979). Originally published as *Enstehung und Entwicklung einer wissenschaftlichen Tatsache: Einführung in die Lehre vom Denkstil und Denkkollektiv* (Basel: Benno Schwabe & Co., 1935), 42.

2. *Encyclopedia Britannica Online*, s.v. "Oxytocin," https://www.britannica.com/science/oxytocin/.

3. *Wikipedia*, s.v. "Oxytocin," https://en.wikipedia.org/wiki/Oxytocin.

4. Henry Hallett Dale, *Adventures in Physiology* (London: Wellcome Trust, 1953), 52. Dale continued to believe that the oxytocic actions of pituitary extracts were properties of the same substance that produced a pressor effect until 1920, when others showed that the pressor and oxytocic actions could be separated.

5. *Wikipedia*, "Oxytocin."

6. H. H. Dale, "The Action of Extracts of the Pituitary Body," *Biochemical Journal* 4 (1909): 427–447.

7. W. B. Bell and P. Hick, "Report CXII. Observations on the Physiology of the Female Genital Organs," *British Medical Journal* 1 (1909): 777–783.

8. *Wikipedia*, "Oxytocin."

9. G. Leng and N. Sabatier, "Measuring Oxytocin and Vasopressin: Bioassays, Immunoassays and Random Numbers," *Journal of Neuroendocrinology* 28, no. 10 (2016), https://doi.org/10.1111/jne.12413.

10. V. Breuil, E. Fontas, R. Chapurlat, P. Panaia-Ferrari, H. B. Yahia, S. Faure, L. Euller-Ziegler, et al. "Oxytocin and Bone Status in Men: Analysis of the MINOS Cohort," *Osteoporosis International* 26 (2015): 2877–2882.

11. O. Weisman, O. Zagoory-Sharon, I. Schneiderman, I. Gordon, and R. Feldman R, "Plasma Oxytocin Distributions in a Large Cohort of Women and Men and Their Gender-Specific Associations with Anxiety," *Psychoneuroendocrinology* 38 (2013): 694–701.

12. Leng and Sabatier, "Measuring Oxytocin and Vasopressin."

13. Abridged and modified, with permission, from G. Leng, "Editorial: The Dogs That Don't Bark," *Journal of Neuroendocrinology* 15 (2003): 1103–1104.

14. Karl Popper, *The Logic of Scientific Discovery* (London: Routledge, 2000 [1959]), 479.

Chapter 15: Organized Skepticism in Science

1. Forster, *Aspects of the Novel*.

2. G. Nave, C. Camerer, and M. McCullough, "Does Oxytocin Increase Trust in Humans? A Critical Review of Research," *Perspectives in Psychological Sciences* 10, no. 6 (2015): 772–789.

3. Nave et al., "Does Oxytocin Increase Trust," 772.

4. H. Walum, I. D. Waldman, and L. J. Young, "Statistical and Methodological Considerations for the Interpretation of Intranasal Oxytocin Studies," *Biological Psychiatry* 79 (2016): 251–257.

5. G. Leng and M. Ludwig, "Intranasal Oxytocin: Myths and Delusions," *Biological Psychiatry* 79 (2016): 243–250.

6. There are exceptions. In medical science, the outcomes of clinical trials are necessarily *probabilistic* rather than *categorical*. Trials cannot be designed to control for confounding factors in the way that laboratory studies can be. People are genetically diverse in ways that laboratory animals are not; they are the products of diverse environments and comply inconsistently with study requirements in multiple ways. Replication of trials is essential to establish whether the same results can be found in different trial settings and with different patient groups. Small trials are commonly unreliable: positive outcomes are often false positives, and the effect sizes that are published in the literature are generally greater than will eventually be found in large studies.

7. G. Leng, S. Maclean, and R. I. Leng, "The Vasopressin–Memory Hypothesis: A Citation Network Analysis of a Debate," *Annals of the New York Academy of Sciences* (2019), in press.

8. W. B. Mens, A. Witter, T. B. van Wimersma Greidanus, "Penetration of Neurohypophyseal Hormones from Plasma into Cerebrospinal Fluid (CSF): Half-times of Disappearance of These Neuropeptides from CSF," *Brain Research* 262 (1983): 143–149.

9. Harry M. Collins, "Stages in the Empirical Programme of Relativism," *Social Studies of Science* 11 (1981): 3–10.

10. Richard Van Noorden, Brendan Maher, and Regina Nuzzo, "The Top 100 Papers," *Nature* 514 (2014): 550–553.

Chapter 16: Webs of Belief

1. J. Secord, "Knowledge in Transit," *Isis* 95 (2004): 654–672.

2. S. A. Greenberg, "How Citation Distortions Create Unfounded Authority: Analysis of a Citation Network," *BMJ* 339 (2009): b2680, 1–14.

3. Greenberg, "How Citation Distortions Create Unfounded Authority," 2.

4. Greenberg, 3.

5. Greenberg, 4.

6. Citation bias is referred to in the literature currently in two senses: (1) a statistically significant difference between the citations to papers reporting results in a particular direction over others reporting results in the opposite direction; (2) one-sided citation bias (or "selective citation") that refers to the complete exclusion of discrepant data. In Greenberg's study, it was largely citation bias of the latter that was detected, but in the following chapter we report studies that have mostly documented the former.

7. S. A. Greenberg, "Understanding Belief Using Citation Networks," *Journal of Evaluation in Clinical Practice* 17 (2011): 389–393.

8. Greenberg, "Understanding Belief Using Citation Networks."

9. R. I. Leng, "A Network Analysis of the Propagation of Evidence Regarding the Effectiveness of Fat-Controlled Diets in the Secondary Prevention of Coronary Heart Disease (CHD): Selective Citation in Reviews," *PLoS ONE* 13, no. 5 (2018): e0197716.

10. L. Trinquart, D. M. Johns, and S. Galea, "Why Do We Think We Know What We Know? A Metaknowledge Analysis of the Salt Controversy," *International Journal of Epidemiology* 45 (2016): 251–260.

11. Trinquart et al., "Why Do We Think We Know," 257.

Chapter 17: Unintended Consequences

1. Francis Bacon, *The Advancement of Learning*, (1605).

2. R. Rosenthal, "The 'File Drawer Problem' and Tolerance for Null Results," *Psychological Bulletin* 86 (1979): 638–641.

3. K. Dickersin, S. Chan, T. C. Chalmers, H. S. Sacks, and H. Smith Jr., "Publication Bias and Clinical Trials," *Controlled Clinical Trials* 8 (1987): 343–353.

4. F. Song, S. Parekh, L. Hooper, Y. K. Loke, J. Ryder, A. J. Sutton, C. Hing, et al., "Dissemination and Publication of Research Findings: An Updated Review of Related Biases," *Health Technology Assessment* 14, no. 8 (2010).

5. M. Kicinski, D. A. Springate, and E. Kontopantelis, "Publication Bias in Meta-analyses from the Cochrane Database of Systematic Reviews," *Statistics in Medicine* 34 (2015): 2781–2793.

6. P. C. Gøtzsche, "Reference Bias in Reports of Drug Trials," *British Medical Journal* 295 (1987): 654–656.

7. U. Ravnskov, "Cholesterol Lowering Trials in Coronary Heart Disease: Frequency of Citation and Outcome," *British Medical Journal* 305 (1992): 15–19.

8. A. S. Jannot, T. Agoritsas, A. Gayet-Ageron, and T. V. Perneger, "Citation Bias Favoring Statistically Significant Studies Was Present in Medical Research," *Journal of Clinical Epidemiology* 66 (2013): 296–301.

9. B. Duyx, M. J. E. Urlings, G. H. M. Swaen, L. M. Bouter, and M. P. Zeegers, "Scientific citations Favor Positive Results: A Systematic Review and Meta-analysis," *Journal of the American Society of Information Science* 88 (2017): 92–101.

10. B. Duyx, M. J. E. Urlings, G. M. H. Swaen, L. M. Bouter, and M. P. Zeegers, "Selective Citation in the Literature on the Hygiene Hypothesis: A Citation Analysis on the Association between Infections and Rhinitis," *BMJ Open* 9 (2019): e026518.

11. B. G. Hutchison, A. D. Oxman, S. Lloyd, "Comprehensiveness and Bias in Reporting Clinical Trials. Study of Reviews of Pneumococcal Vaccine Effectiveness," *Medecin de Famille Canadien* 41 (1995): 1356–1360.

12. Jannot et al., "Citation Bias Favoring Statistically Significant Studies."

13. S. A. Greenberg, "How Citation Distortions Create Unfounded Authority: Analysis of a Citation Network," *BMJ* 339 (2009): b2680.

14. J. Porter and H. Jick, "Addiction Rare in Patients Treated with Narcotics," *New England Journal of Medicine* 302 (1980): 123.

15. P. T. M. Leung, E. M. Macdonald, M. B. Stanbrook, I. A. Dhalla, and D. N. Juurlink, "A 1980 Letter on the Risk of Opioid Addiction," *New England Journal of Medicine* 376 (2017): 2194–2195.

16. Anne-Wil Harzing with Pieter Kroonenberg, "The Mystery of the Phantom Reference," Harzing.com (personal website), October 26, 2017, last updated August 7, 2019, https://harzing.com/publications/white -papers/the-mystery-of-the-phantom-reference.

17. K. A. Robinson and S. N. Goodman, "A Systematic Examination of the Citation of Prior Research in Reports of Randomized, Controlled Trials," *Annals of Internal Medicine* 154 (2011): 50–55.

18. D. L. Sacket, "Bias in Analytic Research," *Journal of Chronic Diseases* 32 (1979): 51–63; R. I. Leng, "A Network Analysis of the Propagation of Evidence Regarding the Effectiveness of Fat-Controlled Diets in the Secondary Prevention of Coronary Heart Disease (CHD): Selective Citation in Reviews," *PLoS ONE* 13, no. 5 (2018): e0197716.

Chapter 18: High-Impact Papers

1. Richard Van Noorden, Brendan Maher, and Regina Nuzzo, "The Top 100 Papers," *Nature* 514 (2014): 550–553, https://doi.org/10.1038/514550a.

2. O. H. Lowry, N. J. Rosebrough, A. L. Farr, and R. J. Randall, "Protein Measurement with the Folin Phenol Reagent," *Journal of Biological Chemistry* 193 (1951): 265–275.

3. R. C. Oldfield, "The Assessment and Analysis of Handedness: The Edinburgh Inventory," *Neuropsychologia* 9 (1971): 97–113.

4. J. P. A. Ioannidis, "Massive Citations to Misleading Methods and Research Tools: Matthew Effect, Quotation Error and Citation Copying," *European Journal of Epidemiology* 33 (2018):1021–1023.

5. A. Stang, "Critical Evaluation of the Newcastle–Ottawa Scale for the Assessment of the Quality of Nonrandomized Studies in Meta-analyses," *European Journal of Epidemiology* 25 (2010): 603–605.

6. A. Stang, S. Jonas, and C. Poole, "Case Study in Major Quotation Errors: A Critical Commentary of the Newcastle–Ottawa Scale," *European Journal of Epidemiology* 33 (2018): 1025–1031.

7. M. V. Simkin and V. P. Roychowdhury, "Read Before You Cite!" *Complex Systems* 14 (2003): 269–274.

8. H. Jergas and C. Baethge, "Quotation Accuracy in Medical Journal Articles—A Systematic Review and Meta-analysis," *PeerJ* 3 (2015): e1364.

9. C. K. Kramer, B. Zinman, and R. Retnakaran, "Are Metabolically Healthy Overweight and Obesity Benign Conditions? A Systematic Review and Meta-analysis," *Annals of Internal Medicine* 159 (2013): 758–769.

10. Global BMI Mortality Collaboration, "Body-Mass Index and All-Cause Mortality: Individual-Participant-Data Meta-analysis of 239 Prospective Studies in Four Continents," *Lancet* 388 (2016): 776–786.

11. D. Berrigan, R. P. Troiano, and B. I. Graubard, "BMI and Mortality: The Limits of Epidemiological Evidence," *Lancet* 388 (2016): 734–6.

12. K. M. Flegal, J. P. A. Ioannidis, and W. Doehner, "Flawed Methods and Inappropriate Conclusions for Health Policy on Overweight and Obesity: The Global BMI Mortality Collaboration Meta-analysis," *Journal of Cachexia Sarcopenia and Muscle* 10 (2019): 9–13.

13. S. W. Yi, H. Ohrr, S. A. Shin, and J. J. Yi, "Sex-Age-Specific Association of Body Mass Index with All-Cause Mortality among 12.8 Million Korean Adults: A Prospective Cohort Study," *International Journal of Epidemiology* 44 (2015): 1696–1705.

14. P. I. Pacy, I. Webster, and I. S. Garrow, "Exercise and Obesity," *Sports Medicine* 3 (1986): 89–113.

15. N. Trefethen, "A New BMI" (letter to the editor), *Economist*, January 5, 2013.

Chapter 19: Are Most Research Published Research Findings False?

1. J. P. A. Ioannidis, "Why Most Published Research Findings Are False," *PLoS Medicine* 2 (2005): e124.

2. N. L. Kerr, "HARKing: Hypothesizing after the Results Are Known," *Personality and Social Psychology Reviews* 2 (1998): 196–217.

3. E. J. Masicampo and D. R. Lalande, "A Peculiar Prevalence of *p* Values Just Below .05," *Quarterly Journal of Experimental Psychology* 65 (2012): 2271–2279.

4. A. Gelman and H. Stern, "The Difference between "Significant" and "Not Significant" Is Not Itself Statistically Significant," *American Statistician* 60 (2006): 328–331.

5. B. C. Martinson, S. A. Anderson, and R. de Vries, "Scientists Behaving Badly," *Nature* 435 (2005): 737–738.

6. H. Nakaoka and I. Inoue, "Meta-analysis of Genetic Association Studies: Methodologies, Between-Study Heterogeneity and Winner's Curse," *Journal of Human Genetics* 54 (2009): 615–623; J. P. A. Ioannidis, T. A. Trikalinos, E. E. Ntzani, and D. G. Contopoulos-Ioannidis, "Genetic Associations in Large versus Small Studies: An Empirical Assessment," *Lancet* 361 (2003): 567–571.

7. K. S. Button, J. P. A. Ioannidis, C. Mokrysz, B. A. Nosek, J. Flint, E. S. Robinson, and M. R. Munafò, "Power Failure: Why Small Sample Size Undermines the Reliability of Neuroscience," *Nature Reviews Neuroscience* 14 (2013): 365–376.

8. F. Mathews, P. J. Johnson, and A. Neil, "You Are What Your Mother Eats: Evidence for Maternal Preconception Diet Influencing Foetal Sex in Humans," *Proceedings of the Royal Society B* 275 (2008): 1661–1668.

9. S. S. Young, H. Bang, and K. Oktay, "Cereal-induced Gender Selection? Most Likely a Multiple Testing False Positive," *Proceedings of the Royal Society B* 276 (2009): 1211–1212.

10. J. S. Cramer and L. H. Lumey, "Maternal Preconception Diet and the Sex Ratio," *Human Biology* 82 (2010): 103–107.

11. J. P. Simmons, L. D. Nelson, and U. Simonsohn, "False-Positive Psychology: Undisclosed Flexibility in Data Collection and Analysis Allows Presenting Anything as Significant," *Psychological Science* 22 (2011): 1359–1366.

12. J. P. A. Ioannidis, "The Importance of Predefined Rules and Pre-specified Statistical Analyses: Do Not Abandon Significance," *JAMA* 321 (2019): 2067–2068.

13. P. Janiaud, I. A. Cristea, and J. P. A. Ioannidis, "Industry-Funded versus Non-Profit-Funded Critical Care Research: A Meta-epidemiological Overview," *Intensive Care Medicine* 44 (2018): 1613–1627.

14. S. Heres, J. Davis, K. Maino, E. Jetzinger, W. Kissling, and S. Leucht, "Why Olanzapine Beats Risperidone, Risperidone Beats Quetiapine, and Quetiapine Beats Olanzapine: An Exploratory Analysis of Head-to-Head Comparison Studies of Second-Generation Antipsychotics," *American Journal of Psychiatry* 163 (2006): 185–194.

15. M. Bes-Rastrollo, M. B. Schulze, M. Ruiz-Canela, and M. A. Martinez-Gonzalez, "Financial Conflicts of Interest and Reporting Bias Regarding the Association between Sugar-Sweetened Beverages and Weight Gain: A Systematic Review of Systematic Reviews," *PLoS Medicine* 10, no. 12 (2013): e1001578.

16. A. Ronald Fisher, *The Design of Experiments* (Edinburgh and London: Oliver and Boyd, 1935), 1–2.

17. D. Trafimow and M. Marks, "Editorial," *Basic and Applied Social Psychology* 37 (2015): 1–2.

18. Ioannidis, "The Importance of Predefined Rules."

Chapter 20: Societal and Economic Impact of Basic Research

1. Gareth Leng, *The Heart of the Brain: The Hypothalamus and Its Hormones* (Cambridge, MA: MIT Press, 2018).

2. M. Widmer, G. Piaggio, T. M. H. Nguyen, A. Osoti, O. O. Owa, S. Misra, A. Coomarasamy, et al. "Heat-Stable Carbetocin versus Oxytocin to Prevent Hemorrhage after Vaginal Birth," *New England Journal of Medicine* 379 (2018): 743–752.

3. Arthur Hugh Clough, "Say not the Struggle nought Availeth."

4. A. Flexner, "The Usefulness of Useless Knowledge," *Harper's Magazine* 179 (1939): 344–352.

5. A. Flexner, *The American College: A Criticism* (New York: The Century Company, 1908).

6. A. Flexner, *Medical Education in the United States and Canada* (Washington, DC: Science and Health Publications, Inc., 1910).

7. Flexner, *Medical Education in the United States and Canada*, 19.

8. Flexner, 55.

9. T. P. Duffy, "The Flexner Report—100 Years Later," *Yale Journal of Biology and Medicine* 84 (2011): 269–276.

10. A. Flexner, *Medical Education: A Comparative Study* (New York: Macmillan, 1925).

11. Health Economics Research Group, Office of Health Economics, and RAND Europe, *Medical Research: What's It Worth? Estimating the Economic Benefits from Medical Research in the United Kingdom* (London: UK Evaluation Forum, 2008).

12. L. Rosenberg, "Exceptional Economic Returns on Investments in Medical Research," *Medical Journal of Australia* 177 (2002): 368–371.

13. Health Economics Research Group, Office of Health Economics, and RAND Europe, *Medical Research*.

14. J. Grant and M. J. Buxton, "Economic Returns to Medical Research Funding," *BMJ Open* 8, no. 9 (2018): e022131.

15. G. Guise, "Margaret Thatcher's Influence on British Science," *Notes and Records of the Royal Society of London* 68 (2014): 301–309.

16. E. Garfield, "Shame on You Mrs Thatcher," *Scientist*, March 9, 1987.

17. R. S. Williams, S. Lotia, A. K. Holloway, and A. R. Pico, "From Scientific Discovery to Cures: Bright Stars Within a Galaxy," *Cell* 163 (2015): 21–23.

18. Williams et al., "From Scientific Discovery to Cures," 22.

19. Flexner, "The Usefulness of Useless Knowledge."

Chapter 21: Lost in Citation

1. V. B. Mahesh and R. B. Greenblatt, "Physiology and Pathogenesis of the Stein–Leventhal Syndrome," *Nature* 191 (1961): 888–890.

2. S. G. Hillier, G. V. Groom, A. R. Boyns, and E. H. D. Cameron, "Development of Polycystic Ovaries in Rats Actively Immunised against T-3-BSA," *Nature* 250 (1974): 433–434.

Chapter 22: Conviction, Expectations, and Uncertainty in Science

1. Karl Popper, *The Logic of Scientific Discovery* (London: Routledge, 2000 [1959]), 280.

2. Stephen Jay Gould, *The Mismeasure of Man* (New York: Norton & Company, 1981).

3. David Joravsky, *The Lysenko Affair* (Chicago: University of Chicago Press, 2010).

4. Christopher Badcock, "The Lasting Lesson of Lysenko," *Psychology Today*, January 11, 2014, https://www.psychologytoday.com/us/blog/the-imprinted-brain/201401/the-lasting-lesson-lysenko.

5. Quoted in P. Bateson, "Haldane at 125: The Cleverest Man I Never Met," *Journal of Genetics* 96 (2017): 801–804.

6. J. B. S. Haldane, "Lysenko and Genetics," *Science and Society* IV, no. 4 (Fall 1940).

7. D. Medina, "Of Mice and Women: A Short History of Mouse Mammary Cancer Research with an Emphasis on the Paradigms Inspired by the Transplantation Method," *Cold Spring Harbor Perspectives in Biology* 2, no. 10 (2010): a004523.

8. J. B. S. Haldane, "Lysenko and Genetics."

9. Y. Liu, B. Li, and Q. Wang, "Science and Politics," *EMBO Reports* 10 (2009): 938–939.

10. Ben Goldacre, *Bad Pharma: How Medicine is Broken, and How We Can Fix It* (New York: HarperCollins, 2013).

11. R. Smith, "Editorial: The Cochrane Collaboration at 20," *BMJ* 347 (2013): f7383.

12. Goldacre, *Bad Pharma*, 1.

13. F. T. Bourgeois, S. Murthy, and K. D. Mandl, "Outcome Reporting among Drug Trials Registered in ClinicalTrials.gov," *Annals of Internal Medicine* 153 (2010): 158–166.

14. Bourgeois et al., "Outcome Reporting among Drug Trials," 159–160.

15. J. Lexchin, L. A. Bero, B. Djulbegovic, and O. Clark, "Pharmaceutical Industry Sponsorship and Research Outcome and Quality: Systematic Review," *British Medical Journal* 326 (2003): 1167.

16. P. H. Thibodeau, R. K. Hendricks, and L. Boroditsky, "How Linguistic Metaphor Scaffolds Reasoning," *Trends in Cognitive Sciences* 21 (2017): 852–863; P. H. Thibodeau and L. Boroditsky, "Metaphors We Think With: The Role of Metaphor in Reasoning," *PLoS ONE* (2011): https://doi.org/10.1371/journal.pone.0016782.

17. K. E. Stanovich and M. E. Toplak, "The Need for Intellectual Diversity in Psychological Science: Our Own Studies of Actively Open-Minded Thinking as a Case Study," *Cognition* 187 (2019): 156–166.

Chapter 23: Journals, Impact Factors, and Their Corrupting Influence on Science

1. Forster, *Aspects of the Novel*.

2. Robert Maxwell (1923–1991) was a business tycoon who built a media empire that included the *Daily Mirror* Group in the United Kingdom and the *New York Daily News*. His flamboyant lifestyle, combined with his ill-fated ventures in newspaper publishing, contributed to increasing

debts, and after his death his empire collapsed into bankruptcy. Obituary in the *Guardian*: https://www.theguardian.com/politics/1991/nov/06/obituaries.

3. S. Buranyi, "Is the Staggeringly Profitable Business of Scientific Publishing Bad for Science?" *Guardian*, June 27, 2017, https://www.theguardian.com/science/2017/jun/27/profitable-business-scientific-publishing-bad-for-science?.

4. RELX Group, "RELX—Results for the Year to December 2018," press release, retrieved April 30, 2019. https://www.relx.com/media/press-releases/year-2019/relx-2018-results.

5. J. D. Watson and F. H. C. Crick, "A Structure for Deoxyribose Nucleic Acid," *Nature* 171 (1953): 737–738.

6. M. E. Falagas and V. G. Alexiou, "The Top-Ten in Journal Impact Factor Manipulation," *Archivum Immunologiae et Therapiae Experimentalis* 56 (2008): 223–226.

7. S. Cantrill, "Imperfect Impact. Chemical Connections," stuartcantrill.com (blog), January 23, 2016, https://stuartcantrill.com/2016/01/23/imperfect-impact/. Cited in D. R. Shanahan, "Auto-correlation of Journal Impact Factor for Consensus Research Reporting Statements: A Cohort Study," *PeerJ* 4 (2016): e1887, https://doi.org/10.7717/peerj.1887.

8. V. Larivière, V. Kiermer, C. J. MacCallum, M. McNutt, M. Patterson, B. Pulverer, S. Swaminathan, S. Taylor, and S. Curry, "A Simple Proposal for the Publication of Journal Citation Distributions," *bioRxiv* (July 5, 2016), https://doi.org/10.1101/062109.

9. G. A. Lozano, V. Larivière, and Y. Gingras, "The Weakening Relationship between the Impact Factor and Papers' Citations in the Digital Age," *Journal of the Association for Information Science and Technology* 63 (2012): 2140–2145.

10. F. C. Fang and A. Casadevall, "Retracted Science and the Retraction Index," *Infection and Immunity* 79 (2011): 3855–3859; F. C. Fang, R. G. Steen, and A. Casadevall, "Misconduct Accounts for the Majority of Retracted Scientific Publications," *Proceedings of the National Academy of Sciences USA* 109 (2012): 17028–17033.

11. B. Brembs, "Prestigious Science Journals Struggle to Reach Even Average Reliability," *Frontiers in Human Neuroscience* 12 (2018): 37, https://doi.org/10.3389/fnhum.2018.00037.

12. A. Molinié and G. Bodenhausen, "Bibliometrics as Weapons of Mass Citation" [*La bibliométrie comme arme de citation massive*], *Chimia* 64 (2010): 78–89.

13. E. Dzeng, "How Academia and Publishing Are Destroying Scientific Innovation: A Conversation with Sydney Brenner," *King's Review*, February 24, 2014, http://kingsreview.co.uk/articles/how-academia-and -publishing-are-destroying-scientific-innovation-a-conversation-with -sydney-brenner/.

14. Jerry Z. Muller, *The Tyranny of Metrics* (Princeton, NJ: Princeton University Press, 2018).

15. Muller, *The Tyranny of Metrics*, 176–177.

16. E. Callaway, "Beat It, Impact Factor! Publishing Elite Turns against Controversial Metric," *Nature* 535 (July 8, 2016): 210–211.

17. "San Francisco Declaration on Research Assessment," https://sfdora .org/.

18. Holly Else, "Radical Open-Access Plan Could Spell End to Journal Subscriptions," *Nature* 561 (September 4, 2018): 17–18; Plan S, making full and immediate open access a reality: https://www.coalition-s.org/.

19. J. Bohannon, "Who's Afraid of Peer Review?" *Science* 342 (2013): 60–65.

20. S. Moore, C. Neylon, M. P. Eve, D. P. O'Donnell, and D. Pattinson, "'Excellence R Us': University Research and the Fetishisation of Excellence," *Palgrave Communications* 3:17010.

21. Molinié and Bodenhausen, "Bibliometrics as Weapons of Mass Citation."

22. R. R. Ernst, "The Follies of Citation Indices and Academic Ranking Lists. A Brief Commentary to 'Bibliometrics as Weapons of Mass Citation,'" *Chimia* 64 (2010): 90, https://doi.org/10.2533/chimia.2010.90.

Chapter 24: The Narrative Fallacy

1. J. J. Ware and M. R. Munafò, "Significance Chasing In Research Practice: Causes, Consequences and Possible Solutions," *Addiction* 110 (2015): 4–8; J. M. Wicherts, "The Weak Spots in Contemporary Science (and How to Fix Them)," *Animals (Basel)* 7, no. 12 (2017), https://doi.org/10.3390/ani7120090; A. Carmona-Bayonas, P. Jimenez-Fonseca, A. Fernández-Somoano, F. Álvarez-Manceñido, E. Castañón, A. Custodio, F. A. de la Peña, R. M. Payo, and L. P. Valiente, "Top Ten Errors of Statistical Analysis in Observational Studies for Cancer Research," *Clinical Translational Oncology* 20 (2018): 954–965; P. De Boeck and M. Jeon, "Perceived Crisis and Reforms: Issues, Explanations, and Remedies," *Psychological Bulletin* 144 (2018): 757–777; P. E. Shrout and J. L. Rodgers, "Psychology, Science, and Knowledge Construction: Broadening Perspectives from the Replication Crisis," *Annual Reviews in Psychology* 69 (2018): 487–510.

2. M. J. Müller, B. Landsberg, and J. Ried, "Fraud in Science: A Plea for a New Culture in Research" *European Journal of Clinical Nutrition* 68 (2014): 411–415.

3. D. Fanelli, "How Many Scientists Fabricate and Falsify Research? A Systematic Review and Meta-analysis of Survey Data," *PLoS ONE* 4, no. 5 (2009): e5738.

4. L. Maggio, T. Dong, E. Driessen, and A. Artino Jr., "Factors Associated with Scientific Misconduct and Questionable Research Practices in Health Professions Education," *Perspectives in Medical Education* 8 (2019): 74–82.

Chapter 25: Scholarship

1. "Minutes of Evidence (10 Jan 2005) of the UK Parliament's Select Committee on Science and Technology," https://publications.parliament.uk/pa/cm200405/cmselect/cmsctech/6/5011003.htm.

2. Karl Popper, *The Logic of Scientific Discovery* (London: Routledge, 2000 [1959]), 280.

3. S. Fortunato, C. T. Bergstrom, K. Börner, J. A. Evans, D. Helbing, S. Milojević, A. M. Petersen, et al., "Science of Science," *Science* 359, no. 6379 (2018).

4. S. Brenner, "Frederick Sanger (1918–2013)," *Science* 343 (2018): 262.

5. D. R. Comer and M. Schwartz, "The Problem of Humiliation in Peer Review," *Ethics and Education* 9 (2014): 141–156.

6. Robert K. Merton, *On the Shoulders of Giants* (New York: Harcourt Brace Jovanovich, 1965).

Index